WHAT IS M.E.?

AN HFME BOOKLET

Essential information on the neurological disease M.E.,
sourced from the world's leading M.E. experts.

FIRST EDITION. PUBLISHED 2012. JODI BASSETT.

BOOKS BY JODI BASSETT

Caring for the M.E. Patient

What is M.E.? (booklet)

COMING SOON FROM THE SAME AUTHOR

Waking from the CFS Misdiagnosis Nightmare (working title)

The HFME Guide to Basic M.E. Treatment (working title)

Overexertion and the M.E. Patient (working title)

What M.E. Feels Like (working title)

M.E.: A Guide for Doctors (working title)

A FOREWORD BY DR BYRON HYDE

Unfortunately, I have not met Jodi Bassett of the Hummingbirds Foundation for M.E. although in September 2010 I came close to visiting her home when I was lecturing in the incredibly beautiful city of Perth in Western Australia. Nor have I met her charming mother or her father who have both been indispensible in assisting Jodi and her work.

Jodi is a very courageous young woman who not only runs one of the few balanced on-line M.E. world newsletters and websites but recently has completed a serious book on the subject of M.E. This is a book that deserves being read, not only by patients and physicians with an interest in M.E. but the bureaucrats in the USA Centers for Disease Control who have done so much damage to the understanding of the M.E. illness spectrum.

It is not just very difficult, it is almost impossible to hold a balanced and intelligent view on a subject such as M.E. without spending one's life actually examining and investigating the patient. Putting the name of a disease on a patient is never enough; what is essential is the proper examination of a disabled patient to understand how as a physician the patient may be assisted.

There is also so much false information that is picked up and disseminated by the M.E. and CFS press it is near impossible to hold one's head above the water and sift through this morass of misinformation. Much of the misinformation on M.E. and CFS is published by individuals and companies who own test laboratories or who hold patents on viruses or viral evaluation. Any attempt to seek the truth is always a major difficulty.

Somehow, Jodi Bassett and Hummingbird have managed to plow through this field of weeds. I recommend her books to all and wish them every best success.

Byron M. Hyde M.D.

Byron Marshall Hyde MD
Sicily, August 2011

DR HYDE WITH HIS WIFE IONE

How to use this book

For friends, family and partners:

- To get the basic facts about M.E. quickly, read one or all of the three brief papers in Chapter 1.
- Chapter 3 is recommended reading where the person you know has severe M.E
- To learn as much as possible about M.E., read the entire booklet.
- The book 'Caring for the M.E. Patient' is also highly recommended for friends, family and partners of M.E. patients.

For carers of M.E. patients:

- To get the basic facts about M.E. quickly, read one or all of the three brief papers in Chapter 1.
- To learn about some of the specific care needs of an M.E. patient, read "Why patients with severe M.E. are housebound and bedbound' on page 50.
- To learn as much as possible about M.E., read the entire booklet.
- The book 'Caring for the M.E. Patient' is also highly recommended for carers.

For doctors:

- Doctors interested in the correct diagnosis, management and treatment of M.E. are advised to read "What is M.E.?" on page 29 and "Why patients with severe M.E. are housebound and bedbound' on page 50.
- Also recommended reading are all of the papers in Chapters 1, 2 and 3 and the book 'Caring for the M.E. Patient.'
- Other relevant papers may be found at the HFME website (www.hfme.org). The most important papers on the HFME website for doctors are: *The misdiagnosis of 'CFS,' M.E. is not fatigue or 'CFS,' Testing for M.E. (and Dr Byron Hyde's 2007 Nightingale definition of M.E.), The myths of M.E., The importance of avoiding overexertion in M.E.* and *Treating M.E. – The basics*

For M.E. patients:

- This booklet makes a very suitable primer for the M.E. patient. Patients may be best served by first reading this book themselves (or as much of it as is possible, beginning with Chapter 1) and then passing it on to friends and family members.
- Additional papers for M.E. patients can be found at the HFME website (www.hfme.org).
- The book 'Caring for the M.E. Patient' is also highly recommended for M.E. patients.

Contents

Notes on the formatting, design & use of this book

Before reading this book, please note:

1. As some visitors to the HFME website (www.hfme.org) or readers of this book may only ever read one HFME paper, each paper has been designed to be a stand-alone resource which focuses on one aspect of Myalgic Encephalomyelitis (M.E.) but which also includes a brief rundown of the basic facts of M.E. Thus there is significant repetition of the basic facts of M.E. (and related topics) from paper to paper in this book.

If you have read the basic facts once and have no need to be reminded of this information again, please just skim over the repetitive sections when you encounter them in future papers.

2. The papers in this book were originally created to be published online, and distributed for free, on the HFME website. There are many small differences in how information is presented online and in print form. In an ideal world each HFME paper would have been completely reformatted and reorganised, before being included in this book. Unfortunately, due to the serious illness and disability suffered by the author/s, total reformatting and reorganising of each paper was not possible.

Thus this book includes some minor formatting inconsistencies. Where further information is recommended, the links given are in an online format (i.e. HTML links appear here as underlined text). There are also almost certainly some minor grammatical errors.

However, we ask readers to ignore these superficial imperfections and to focus on the far more important fact that the information given in this book on M.E. (and related topics) is rock solid. It has been compiled using information from the world's leading M.E. experts – and a large number of M.E. patient accounts spanning many decades – and is of the highest quality. This is information that is currently unknown by most of the public, the media, doctors and even patients themselves, and that desperately needs to become known – this is why those involved with HFME have produced this book, despite their serious illness and disability caused by M.E.

3. To be able to follow any of the 'links' to further information given in this book (represented by underlined text), just go to the HFME website, view the online version of the relevant paper, and click on the relevant link.

4. If you would like printouts of any of the papers in this book for yourself or to hand out to doctors or others, you can download free printable copies of each paper in this book from the HFME website. See the 'Document downloads' page on the website for more information.

Permission is given for all HFME papers in this book to be freely redistributed by email or in print for any not-for-profit purpose provided that the entire text (including the copyright notice, the author's attribution and the HFME logo) is reproduced in full and without alteration. Knowledge is power. Please redistribute these texts, and the HFME books, widely.

5. To learn more about HFME and the aims of HFME, please see Chapter 4 of this book.

6. To read the full reference list for each HFME paper, please see Chapter 4 of this book.

CHAPTER ONE
Introduction

The basic facts of Myalgic Encephalomyelitis (M.E.) cannot be explained in a mere two or three sentences, but these facts are not time consuming to take in or impossibly difficult to understand. This is particularly true when M.E. is separated out from the vague, illogical and designed-to-be-confusing mess that is 'CFS' (Chronic Fatigue Syndrome).

This book will start with the basics, which are:

- To describe briefly what M.E. is,
- To explain some of the terrible problems that people with M.E. are facing, and why they so badly need your support and understanding,
- To provide a brief explanation of *why* these problems are happening and why they continue to occur.

These fundamental facts and issues will help you better understand M.E. and those who suffer from it.

This introductory chapter includes the following three papers:

1. The one page summary of the facts of M.E.

2. M.E.: The shocking disease.

3. What is M.E.? Summary

A one-page summary of the facts of M.E.

- Myalgic Encephalomyelitis (M.E.) is a disabling neurological disease that is very similar to Multiple Sclerosis (M.S.) and Poliomyelitis. Earlier names for M.E. were 'atypical Multiple Sclerosis' and 'atypical Polio.'

- M.E. is a neurological disease characterised by scientifically measurable post-encephalitic damage to the brain stem. This damage is an essential part of M.E., hence the name M.E. The term M.E. was coined in 1956 and means: my = muscle, algic = pain, encephalo = brain, mye = spinal cord, tis = inflammation. This neurological damage has been confirmed in autopsies of M.E. patients.

- Myalgic Encephalomyelitis has been recognised by the World Health Organisation's International Classification of Diseases since 1969 as a distinct organic neurological disease. M.E. is classified in the current WHO International Classification of Diseases with the neurological code G.93.3.

- M.E. is primarily neurological, but also involves cognitive, cardiac, cardiovascular, immunological, endocrinological, metabolic, respiratory, hormonal, gastrointestinal and musculo-skeletal dysfunctions and damage. M.E. affects all vital bodily systems and causes an inability to maintain bodily homeostasis. More than 64 individual symptoms of M.E. have been scientifically documented.

- M.E. is an acute (sudden) onset, infectious neurological disease caused by a virus (a virus with a 4-7 day incubation period). M.E. occurs in epidemics as well as sporadically and over 60 M.E. outbreaks have been recorded worldwide since 1934. There is ample evidence that M.E. is caused by the same type of virus that causes Polio; an enterovirus.

- M.E. can be more disabling than M.S. or Polio, and many other serious diseases. M.E. is one of the most disabling diseases that exists. More than 30% of M.E. patients are housebound, wheelchair-reliant and/or bedbound and are severely limited with even basic movement and communication.

- *Why are M.E. patients so severely and uniquely disabled?* For a person to stay alive, the heart must pump a certain base-level amount of blood. Every time a person is active, this increases the amount of blood the heart needs to pump. Every movement made or second spent upright, every word spoken, every thought thought, every word read or noise heard requires that more blood must be pumped by the heart.

  However, the hearts of M.E. patients only barely pump enough blood for them to stay alive. Their circulating blood volume is reduced by up to 50%. Thus M.E. patients are severely limited in physical, cognitive and orthostatic (being upright) exertion and sensory input.

  This problem of reduced circulating blood volume, leading to cardiac insufficiency, is why every brief period spent walking or sitting, every conversation and every exposure to light or noise can affect M.E. patients so profoundly. Seemingly minor 'activities' can cause significantly increased symptom severity and/or disability (often with a 48-72 hour delay in onset), prolonged relapse lasting months, years or longer, permanent bodily damage (e.g. heart damage or organ failure), disease progression or death.

  If activity levels exceed cardiac output by even 1%, death occurs. Thus the activity levels of M.E. patients must remain strictly within the limits of their reduced cardiac output just in order for them to stay alive. *M.E. patients who are able to rest appropriately and avoid severe or prolonged overexertion have repeatedly been shown to have the most positive long-term prognosis.*

- M.E. is a testable and scientifically measurable disease with several unique features that is not difficult to diagnose (within just a few weeks of onset) using a series of objective tests (e.g. MRI and SPECT brain scans). Abnormalities are also visible on physical exam in M.E.

- M.E. is a long-term/lifelong neurological disease that affects more than one million adults and children worldwide. In some cases M.E. is fatal. (Causes of death in M.E. include heart failure.)

M.E. - The shocking disease

In thinking about M.E. and all of the terrible things that are happening so unfairly to so many wonderful innocent people year after year, and how extremely severe a disease it can be physically, I keep coming back to one word: shocking. These are the basic M.E. facts:

- M.E. is similar in significant ways to illnesses such as Multiple Sclerosis (M.S.), Lupus and Polio.

- M.E. occurs in epidemic and sporadic forms, over 60 M.E. outbreaks have been recorded worldwide since 1934.

- What defines M.E. is a specific type of acquired damage to the brain (the central nervous system) caused by a virus (an enterovirus). It is an *acute (sudden) onset* neurological disease initiated by a virus infection with multi system involvement which is characterised by post encephalitic damage to the brain stem.

- The term M.E. was coined in 1956 and means: My = muscle, algic = pain, encephalo = brain, mye = spinal cord, itis = inflammation. This neurological damage has been confirmed in autopsies of M.E. patients.

- M.E.is primarily neurological, but also involves cognitive, cardiac, cardiovascular, immunological, metabolic, respiratory, hormonal, gastrointestinal and musculo-skeletal dysfunctions and damage. M.E. causes an inability to maintain bodily homeostasis. More than 64 individual symptoms of M.E. have been scientifically documented.

- M.E. can be more disabling than M.S. or Polio, and many other serious diseases. M.E. is one of the most disabling diseases there is. More than 30% of M.E. patients are housebound, wheelchair-reliant and/or bedbound and are severely limited with even basic movement and communication. In some cases M.E.is fatal.

- The hearts of M.E. patients barely pump enough blood for them to stay alive. Their circulating blood volume is reduced by up to 50%. Thus M.E. patients are severely limited in physical, cognitive and orthostatic (being upright) exertion and sensory input. This problem of reduced circulating blood volume, leading to cardiac insufficiency, is why every brief period spent walking or sitting, every conversation and every exposure to light or noise can affect M.E. patients so profoundly.

  Seemingly minor 'activities' can cause significantly increased symptom severity and/or disability (often with a 48-72 hour delay in onset), prolonged relapse lasting months, years or longer, permanent bodily damage (e.g. heart damage or organ failure), disease progression or death.

  If activity levels exceed cardiac output by even 1%, death occurs. Thus the activity levels of M.E. patients must remain strictly within the limits of their reduced cardiac output just in order for them to stay alive.

  M.E. patients who are able to rest appropriately and avoid severe or prolonged overexertion have repeatedly been shown to have the most positive long-term prognosis.

- M.E.is a testable and scientifically measurable disease with several unique features that is not difficult to diagnose, even within just a few weeks of onset, using a series of objective tests.

- M.E.is a debilitating neurological disease which has been recognised by the World Health Organisation (WHO) since 1969 as a distinct organic neurological disorder. M.E. is classified in the current WHO International Classification of Diseases with the neurological code G.93.3.

- Many patients with M.E. do not have access to even basic appropriate medical care. Medical abuse and neglect is also extremely common and often results in the disease becoming severe (and in some cases, death is caused).

- Governments around the world are currently spending $0 a year on M.E. research.

These facts however, fall far short of getting across what a hell on earth M.E. really is. Above all else, I think M.E. is a shocking disease. These are a few of the biggest shocks I've faced, and that others with M.E. also experience:

1. The shock of extremely severe sudden illness and disability

The first big shock is how quickly and completely your entire life can change forever. Having your body suddenly act very differently and not be able to do all the things you have done many thousands or millions of times before, is surreal. This is especially so when this occurs suddenly from one day to the next, as it does with M.E. The sense of unreality can be so strong that you almost wonder why everyone else is still going on as if nothing had changed and everything was normal.

For me, in March 1995 at the age of 19, I went from being very healthy and happy one day to having problems standing upright for more than a few minutes at a time, the next. I also suddenly had severe problems sleeping, thinking and remembering, speaking and understanding speech, eating many foods that I previously tolerated perfectly well, coping with even low levels of noise and light and vibration, coping with warm weather, sitting, with my heart and blood pressure, with any type of physical or cognitive activity causing severe relapse unless within very strict limits as well as memory loss, facial agnosia, learning difficulties, severe pain, alcohol intolerance, blackouts and seizures, intense unusual headaches, burning eyes and ear pain, rashes (and other skin problems), severe nausea and vertigo, total loss of balance when I closed my eyes or the room was dark, muscle weakness and paralysis, and so on.

I suddenly had over 60 individual symptoms, and could only do 40% or less of my pre-illness activities.

It's a bit like one day waking up and suddenly everyone around you is speaking another language and looking at you strangely for not being able to understand what is being said.

At first, not only is it very hard to just accept, but also to really believe it is happening, and that it won't just go away as suddenly as it came. It's all just such a big shock.

2. The medical system shock

As if that weren't enough all on its own, the next big shock involves lifelong beliefs about our medical system. You soon find out that the disease you have is one of those that is treated differently from many others, that not every disease is viewed equally, and that bizarrely this has *nothing at all* to do with the type of disease or the severity of the disease or its symptoms, or testable abnormalities, or the possibility of death, but other non-scientific and non-medical factors. It has to do with political and financial factors, and marketing.

You find out that some diseases get you 'red carpet' treatment or and guarantee that you are treated very well, others are treated adequately, and unfortunately several leave you with no real care at all. Even worse, some diagnoses subject you to serious mistreatment from the professionals *meant to be there to help you*.

Most people trust absolutely that if they get severely ill, they can go to an emergency room and be given the appropriate medical care. I used to trust in that too. But I was to soon find out the hard way that that didn't apply to me anymore. If I went to the emergency room, there was an enormous chance I'd not only get no help at all, but be ridiculed into the bargain or told 'to stop exaggerating' or refused the appropriate tests (and have older test results ignored). I may then be told, illogically, and despite all the evidence to the contrary that 'there is nothing wrong with you, go home and let us care for someone who is really ill.' I'd be far more likely to come out of the emergency room far sicker than when I'd gone in (in crisis), as well as being verbally abused and insulted as well.

Dealing with GPs and specialists is much the same most of the time, for those with M.E. Probably the most common treatment recommended to patients with M.E. is graded exercise therapy (GET) (both formal and informal programs). This is a 'treatment' that can and very often does leave M.E. patients, including children, far sicker afterwards for months, years or longer (wheelchair-reliant, bedbound, needing intensive care etc.). It can also cause death. While it may help some of those with other illnesses very different to M.E., it has a zero percent chance of providing any benefit to M.E. patients. If even a tiny percentage, say 2%, of almost any other patient group were made as ill and disabled by any treatment (as M.E. patients are by GET) it would be a huge scandal. It would make all the papers and there would be all sorts of legal actions and enquiries, and outpourings of public outrage. Yet the incidence of M.E. patients being recommended, or forced or coerced, into this torture is growing every year. Nobody much cares or even knows. It's more than shocking or just very cruel, it's obscene.

Most people have no idea that all this medical abuse occurs regularly, to people *just as ill or even far more ill* as those with M.S. or Lupus. When you do try to tell them, most often they refuse to believe it could be true, so strong is their belief in the fairness and logic of our health system and how much thought, objectivity and careful investigation supposedly goes into giving a final diagnosis and recommending treatment. It's a shocking loss, this loss of belief in a health safety net and a medical system based on logic, science and due care. It's such a comforting belief, it's hardly surprising people don't want to give it up, even if it is false.

Thanks to inappropriate medical care, I, as with many other M.E. patients, soon struggled to do even 5% of the activities I had pre-illness. I was made housebound and 99% bedbound, and have remained so for the last 10 years. My heart-rate skyrockets and my blood pressure drops dramatically after just a few minutes of standing or other overexerting activity. It feels like a heart attack in every organ, and as if my heart is about to explode, or just stop. (The highest heart-rate measurement I've had is 170 bpm and the lowest blood pressure measurement is 79/59 – both were taken at times when I was only moderately ill, relatively speaking, nowhere near my most severe state. Scary.)

I have spent most of the last decade, alone and in pain in a dark quiet room, coping with many different and hideous symptoms. I accept that some people get ill, and that I am at risk of this as much as anyone. What is hard to take is that, like so many M.E. patients, my reaching such a severe disability level and losing so much of my life was completely unnecessary and would very likely not have happened had I had even the most basic appropriate support in the beginning.

3. The welfare system shock

Despite being extremely ill and disabled, M.E. patients are often shocked to find that getting the basic welfare payments is very difficult or impossible. Bizarrely enough, the system is set up in such a way that you can actually be *too ill* to qualify, as so many hoops are required to be jumped through to lodge a successful claim, without which the claim is denied. Ironically, the government agencies seem to have little interest in this conundrum, nor in how much sicker jumping through all their hoops makes you long-term. The ignorance of doctors and their inability to give you an unbiased examination is also a huge problem.

Again, what is far more important to them is the name and reputation of your disease, not how ill and disabled you are. It is not uncommon to find instances of M.E. patients living for years with no disability payments, having to live on the mercy of family, or becoming homeless.

4. The media shock

The public largely trusts the information given about different diseases in the media. I did too, and I still do largely, provided the article is about M.S. or cancer. But like many M.E. patients, I was shocked to find out that when it came to diseases like mine, there was no onus at all on the reporter to be accurate. While a furore would ensure if articles made up *entirely* of false information were printed about M.S. or cancer, almost every article that I read about M.E. was of an unbelievably low standard, yet nobody seemed to care at all.

Similarly, the outrage when certain groups are made fun of in what is deemed an offensive manner, simply does not occur when it's M.E. that is being ridiculed. For some reason M.E. patients (in the UK particularly) are fair game. This is because despite the fact that our governments have created laws designed to stop discrimination on the basis of gender, race and disability and so on, discrimination against M.E. patients is not only allowed, but is actively supported and promoted by government. (For information on *why* this occurs, see <u>What is M.E.?</u>)

5. The human rights groups shock

While the big human rights groups seem very eager to help many other groups and even individuals facing small or large problems, they seem completely unwilling to even look at the severe abuse of human rights facing perhaps a million M.E. patients worldwide, let alone do anything at all to actually help. This when even the smallest action on their part, the smallest indication of support for the M.E. cause, would be such a huge step forward for M.E. patient rights. Such ignorance and injustice is shocking.

6. The friends and family shock

What makes coping with all these things unimaginably worse is having to do so with little if any support from friends and family – and even while facing abuse or ridicule from friends and family. Some patients are even disowned by their whole family, or all but a few members.

Loved ones often believe the same financially-motivated media and government-sanctioned nonsense about your disease as the doctors do. They often accept the 'miracle cure' stories in the media featuring people with a hundred different mild (and sometimes psychological) or transient diseases jumping up and down about how they have been 'cured' by the mumbo jumbo money-making scam of the week – despite the fact that none of these stories features actual M.E. patients, or even patients with diseases similar to M.E.

It's such a huge shock that those you love could see you so ill and refuse to support you and that they have more trust in doctors than in your integrity. They can't believe that if you were seriously ill, a doctor could miss it, even though that is exactly what has happened. They can't believe that the media would be allowed to print completely fictional information about your disease often based on mixed and *entirely unrelated* patient groups, even though they *are* doing just that. Not having medical or media (or government) support makes getting support from loved ones almost impossible.

Having loved ones not stand by you hurts a lot, in many ways. It takes yet another huge swipe at what self-esteem you have left after being treated like dirt by your trusted doctors and welfare departments, leaving study incomplete and/or losing your job and your ability to support yourself and/or being denied the services of a carer when you urgently need one. After so many attacks on your integrity and worth you can't help but be worn down by it all, particularly when you're so ill and even more so if you are not yet of adult age when you become ill. You inevitably feel, not depressed, but as if you must personally be unworthy somehow of any type of care or compassion. Such messages inevitably sink in to some extent after constant repetition, no matter how educated, strong or mentally fit you are.

7. The M.E. charities and support groups shock

Realising that very nearly all of the charities and support groups that claim to be there to help you actually do not represent or support you at all and are actually hostile to your interests is yet another huge shock.

You go to a group that you trust finally to give you the unadulterated facts and to be working towards improving things and all you get is more abuse and misinformation. Just as bad, you also don't get all the important information about M.E. that could make an enormous positive difference to your life and to your health. If you try to improve matters and provide these groups and individuals with accurate information you are either ignored or banned, perceived as negative.

M.E. patients are in a terrible position. Almost all 'our' charities have sold themselves off to the highest bidder, and are now working to promote the same harmful misinformation they were created to fight against.

(The concepts of 'CFS' and 'ME/CFS' can be immensely profitable to some groups, as is explained in several other HFME papers.) These groups claim to be representing a large and diverse patient group but in reality they do not work for the benefit of any group, except themselves. They often take advantage of patients' lack of ability (or unwillingness) to engage with politics, to read and assimilate significant amounts of slightly complex text and of their goodwill and trust, in the cruellest way. Many patients put all their faith and efforts into this false advocacy, led by vested interest groups. Many (perhaps even most) fellow patients are, unwittingly, working directly against their own interests and aiding their abusers. Many seem determined to support the same old illogical nonsense that is the reason that no progress at all has been made in over 20 years. Perhaps some patients are too ill to even investigate other sources of information than the charity, or they have taken the charity's word for it that the (entirely bogus) information they provide is all that exists.

These sell-out groups and individuals are at fault here to a large extent, but at the same time they couldn't keep doing the evil things they do if they didn't have so much undeserved (and extremely unwise) patient support. It's so incredibly shocking, and *frustrating.*

Those few groups and individuals that are involved in genuine advocacy are often able to do very little due to the physical constraints of M.E., the poverty associated with M.E., and the lack of public and other support. M.E. patients are just too ill to fight effectively for themselves like AIDS patients did. They can't rally or march and many can only barely read or write now. AIDS patients also often have an early asymptomatic period of illness, which enables significant contribution to activism – but for M.E. patients the severe symptoms begin on day one.

8. The M.E. advocacy nightmare shock

Perhaps most shockingly of all, when you try to do some advocacy yourself and tell people about the double standards, discrimination and unfair treatment, and show them mountains of solid facts, you are met with disbelief. People cannot or will not believe that doctors could be so cruel, unscientific, ignorant and illogical; or that our governments and media could be so unethical and dishonest by selling their integrity for political and financial gain; or that so-called 'charities' could be just as corrupt.

Many people refuse to even do a tiny bit of quality reading on the topic of M.E., wrongly believing that they already have all the facts and know all there is to know, believing that anything that they don't know just can't be true. If you try to give people correct information you are accused of exaggerating or being fanciful. People snicker or roll their eyes when you talk about cover-ups, and give your information as much credence as stories of alien abductions or the 'false' moon landing. Anything not already mainstream is met with scepticism, as is the idea that all of these groups could *collaborate* to create a mutually profitable, and very hard to undermine, lie.

Despite ample evidence of similar scandals and cover-ups in the past, people seem unwilling to give up their belief in a fair and just government, media and medical system. They refuse to give up their comforting delusions...until and unless something similar happens to them, at least, and they have no choice but to face reality. But then, of course, they too are disbelieved when they try to spread the word, and so on and on it goes.

Most families and friends of patients are completely unwilling to help with advocacy, very often due to ignorance about the medical and political facts of M.E. Others are too busy with the duties of a carer for advocacy. Patients with other diseases almost always do not understand that the most commonly given information on M.E. is entirely false. By believing M.E. is something it is not and reinforcing many of the worst myths about the disease, most of these well-meaning groups and individuals work directly against the interests of M.E. patients, sadly.

M.E. itself also seems to work against you, in an unexpected way. People say it's too severe and there are too many symptoms. The entirely unique way we respond to even trivial exertion and are so disabled by it, instead of inspiring sympathy, seems to actually inspire disbelief. People seem to (bizarrely) believe that there must be some limit on how bad a disease could be, and that such severe illness couldn't be possible long-term. That you couldn't possibly be too ill to sit or stand up, use the phone, speak or be spoken to, listen to music, write a letter, spend time in hospital or take a short trip out of the house; that you couldn't possibly be so ill that you can only dream of one day being *well enough* to use an electric wheelchair sometimes, if you're really lucky – and so on. As if all humans were 'guaranteed' somehow to always be able to at least do such simple tasks, and to only ever suffer a 'reasonable' level or time period of disability. But the body does not acknowledge such limits. If only.

In 20 years not only has no progress been made in the fight for basic rights, but things have become much worse for M.E. patients and they continue to grow worse still as the years pass.

M.E. is a shocking disease in every way.

M.E. is at least as disabling as any of the other very serious diseases (such as M.S.) and the extremely high level of suffering and isolation it causes can last for many years or decades at a time. Yet M.E. patients get the least amount of support and compassion and such high levels of abuse and outright *ridicule*.

Some of us have some family and/or friends on board, some have welfare, some have basic medical care (although almost none have the same level of care the average M.S. patient has). But most don't have all or even most of these things and when they do they have often taken many years to get and are very hard won.

By the time many of us have some of these things we have been made severely ill by going so long without the right care, that it's a somewhat hollow victory. Especially when we also know that so many others aren't so lucky and that every year thousands of patients, adults as well as teenagers and very young children, are still needlessly being made severely ill or dead through ignorance and misinformation.

It's like an episode of 'The Twilight Zone.' You want to wake up and scream some mornings, thinking it's a nightmare and that such a hell just couldn't possibly be real. That so many innocent people could be so ill, abused and persecuted, with almost none of the public even caring or knowing. That such a flimsy and unethical global medical scam couldn't be so successful at fooling almost everyone, despite the fact it's based on nothing more than smoke and mirrors, scientifically speaking. It's all just far too shocking to take in sometimes.

I invite readers to be shocked about what is happening, even if M.E. hasn't yet affected someone you love or know. The facts are profoundly shocking – I haven't explained even half of them here.

If you have the facts about M.E. you should be not only shocked by what is happening, but also appalled, disgusted and outraged. I beg you to please use that shock, act on it and use it to help try and change things, and to see M.E. patients finally get some basic fair treatment and justice.

The only way change will occur is through education, with enough people simply refusing to accept what is happening anymore.

M.E. patients need your help so desperately, right now. Thank you for taking the time to read this paper.

To read a fully-referenced version of the medical information in this text compiled using information from the world's leading M.E. experts, please see the 'What is M.E.?' paper in this book or on the HFME website.

Acknowledgments
Thanks to Emma Searle, Lesley Ben and Peter Bassett for editing this paper.

Relevant quotes
'Myalgic Encephalomyelitis (M.E.) is distinguished by a unique clinical and epidemiological pattern characteristic of enteroviral infection. It has an UNIQUE neuro-hormonal profile.'
DR ELIZABETH DOWSETT

What is M.E.? Summary

 Myalgic Encephalomyelitis (M.E.) is a debilitating acquired neurological disease that has been recognised by the World Health Organisation (WHO) since 1969 as a distinct organic neurological disorder.

M.E. can occur in both epidemic and sporadic forms; over 60 outbreaks of M.E. have been recorded worldwide since 1934.

M.E. is similar in a number of significant ways to Multiple Sclerosis, Lupus and Poliomyelitis (Polio). It can become extremely severe and disabling and in some cases is fatal.

Is M.E. a new illness?

No. The illness has been documented as an organic (physical) neurological disease for centuries.

The name Myalgic Encephalomyelitis was coined in 1956 in the UK.

M.E. has nothing to do with 'fatigue'

Unlike 'Chronic Fatigue Syndrome' (CFS) M.E. is a neurological illness of extraordinarily incapacitating dimensions that affects virtually every bodily system. Fatigue is not a defining (or essential) symptom of M.E. M.E. and 'CFS' are not at all the same thing.

Why do some groups claim that M.E. and 'CFS' are synonymous terms?

This new name and case definition of 'CFS' was created in the United States by a board of 18 members, few of which had either looked at an epidemic of M.E. or examined *any* patients with the illness.

Why? Money! In the late 1970s and 1980s there was an enormous rise in the reported incidence of M.E. causing alarm among American medical insurance companies.

It was at this time when, in order to side-step the financial responsibility of the many new incoming claims, those involved in the medical insurance industry (on both sides of the Atlantic) began their campaign to reclassify this severely incapacitating and discrete neurological illness as a psychological or 'personality' disorder.

As Professor Malcolm Hooper explains:

> A political decision was taken to rename M.E. as "CFS", the cardinal feature of which was to be chronic or on going "fatigue", a symptom so universal that any insurance claim based on "tiredness" could be expediently denied. The new case definition bore little relation to M.E.: objections were raised by experienced international clinicians, but all objections were ignored.

Public, medical and governmental understanding of M.E. is a huge mess, that is for certain – but it is not an *accidental* mess. (For more information see: <u>Who benefits from 'CFS' and 'ME/CFS'?</u>)

What does a diagnosis of 'CFS' actually mean?

Those diagnosed using the flawed 'CFS' definitions are from a heterogeneous (mixed) population with various misdiagnosed psychiatric and miscellaneous non-psychiatric states that have little in common except the symptom of fatigue. 'CFS' is a wastebasket diagnosis; a mere diagnosis of exclusion.

The fact that a person qualifies for a diagnosis of 'CFS' based on any of the 'CFS' definitions (a) does not mean the patient has M.E., and (b) does not mean she or he has any other distinct and specific illness named 'CFS.' A diagnosis of 'CFS' – based on any of the 'CFS' definitions – can only ever be a *mis*diagnosis.

What is M.E.? What is its symptomatology?

M.E. is characterised primarily by damage to the central nervous system (the brain) initiated by an enteroviral infection that results in dysfunctions and damage to many of the body's vital systems as well as a loss of normal internal homeostasis.

M.E. symptoms are manifested by virtually all bodily systems including: cognitive, cardiac, cardiovascular, immunological, endocrinological, respiratory, hormonal, gastrointestinal and musculo-skeletal dysfunctions and damage. These symptoms are exacerbated by physical and cognitive activity, sensory input and orthostatic stress beyond the individual's limits. In addition to the risk of relapse, repeated or severe overexertion can also cause permanent damage (e.g. to the heart), disease progression and/or death. Symptoms of M.E. include:

> Sore throat, chills, sweats, low body temperature, low grade fever, lymphadenopathy, muscle weakness (or paralysis), muscle pain, muscle twitches or spasms, hair loss, nausea, vomiting, vertigo, cardiac arrhythmia, orthostatic tachycardia, orthostatic fainting or faintness, photophobia and other visual and neurological disturbances, hyperacusis, alcohol intolerance, gastrointestinal and digestive disturbances, allergies and sensitivities to many previously well-tolerated foods, drug sensitivities, stroke-like episodes, nystagmus, difficulty swallowing, myoclonus, temporal lobe and other types of seizures, an inability to maintain consciousness for more than short periods at a time breathing difficulties, emotional lability and sleep disorders.

> Cognitive dysfunction may be pronounced and can include: difficulty/loss of ability in speaking or understanding speech; difficulty in reading, writing or performing basic mathematical tasks as well as having problems with memory including difficulty making new memories and recalling formed memories; difficulties with visual and verbal recall.

What does cause M.E.? Are there outbreaks?

A review of early outbreaks in the history of M.E. shows clinical symptoms were consistent in over 60 recorded epidemics spread all over the world as far back as 1934. M.E. is an acutely acquired neurological illness initiated by a viral (enteroviral) infection with a 4-7 day incubation period. This point of view is supported by history, incidence, symptoms and similarities with other viral illnesses as well as a large body of research spanning decades.

So what do we know about M.E. so far?

There is an abundance of research that shows M.E. is an organic illness that can have profound effects on many bodily systems. Many aspects of the pathophysiology of the disease have been medically explained, and to date there are volumes of articles written, from which more than a thousand good articles support the basic premise of M.E. While there is yet no *single* laboratory test able to diagnose M.E., there *are* a specific *series of tests* which enable an M.E. diagnosis to be easily confirmed; i.e. MRI and SPECT scans of the brain.

Some of the abnormalities found in M.E. patients include: extremely low circulating blood volume (up to an astounding 50%), enzyme pathway disruptions, punctate lesions in M.E. brains resembling those of Multiple

Sclerosis; sub-optimal cardiac function and abnormal cardiovascular responses; persistent viral infection in the heart, severe mitochondrial defects and significantly reduced lung functioning.

Also, strong evidence exists to show (even mild or moderate) exercise can have extremely harmful effects on M.E. patients; permanent damage may be caused as well as disease progression and even death. For this reason, danger exists when medical professionals recommend (and sometimes insist on or even *force*) M.E. patients, including children, to partake in exercise as a treatment to their diagnosis of 'CFS.' Under these harmful circumstances, the M.E. patient is undergoing what amounts to actual legalised torture. Patient accounts of exiting exercise programs much more severely ill than when they entered them, *being wheelchair-bound, bed-bound or needing intensive care* are common. *Deaths have also been reported in M.E. patients following exercise.*

How common is M.E. and who gets it?

M.E. has a similar strike rate to Multiple Sclerosis. M.E. affects more than one million children as young as five, as well as teenagers and adults. It affects all ethnic and socio-economic groups, and has been diagnosed all over the world.

Recovery from and severity of M.E.

M.E. can be progressive, degenerative (change of tissue to a lower or less functioning form, as in heart failure), chronic, or relapsing and remitting. It can also be fatal. Patients who are given advice to rest in the early stages of the illness (and who avoid overexertion thereafter) have repeatedly been shown to have the most positive long-term prognosis. M.E. is a life-long disability where relapse is always possible. Symptoms are extremely severe for at least 30% of sufferers leaving many of them housebound, bedbound and severely disabled.

Truly M.E. can be one of the most devastating and horrific illness there is, yet many with M.E. are subject to repeated medical abuse and neglect because of the way the illness has been dishonestly 'marketed' to the public as being psychological or 'behavioural,' or as being a problem of mere 'fatigue' or a 'fatigue syndrome.'

Sub-grouping or refining or renaming 'CFS' will only waste another 20 years. *There is no such distinct disease/s as 'CFS.'* For the benefit of all the patient groups involved, the bogus disease category of 'CFS' must be abandoned and patients with M.E. must again be diagnosed with M.E. and treated for M.E.

Due to an overwhelming amount of compelling scientific evidence, in 1969 the World Health Organization correctly classified M.E. as a distinct organic neurological disease. This classification/definition and name must be accepted and adhered to in all official documentations and government policy.

PLEASE help to spread the truth about Myalgic Encephalomyelitis.

This appalling abuse and neglect of so many severely ill and vulnerable people on such an industrial scale is inhumane and has already gone on far too long. This will only change through education. People with M.E. desperately need your help.

Acknowledgments
Thanks to Roseanne Schoof and Emma Searle for editing this paper.

Relevant quotes
'People in positions of power are misusing that power against sick people and are using it to further their

own vested interests. No-one in authority is listening, at least not until they themselves or their own family join the ranks of the persecuted, when they too come up against a wall of utter indifference.'
PROFESSOR M. HOOPER 2003

'Central nervous system dysfunction, and in particular, inconsistent CNS dysfunction is, undoubtedly both the chief cause of disability in M.E. and the most critical in the definition of the entire disease process. Of the CNS dysfunctions, cognitive dysfunction is one of the most disabling characteristics of ME. When this simple fact is understood, it become immediately apparent why this is such a devastating disease for children, students and adults, both within and outside the educational system.'
DR BYRON HYDE AND ANIL JAIN MD IN 'THE CLINICAL AND SCIENTIFIC BASIS OF M.E. P 43

'The vested interests of the Insurance companies and their advisers must be totally removed from all aspects of benefit assessments. There must be a proper recognition that these subverted processes have worked greatly to the disadvantage of people suffering from a major organic illness that requires essential support of which the easiest to provide is financial. The poverty and isolation to which many people have been reduced by M.E. is a scandal and obscenity.'
PROFESSOR MALCOLM HOOPER 2006

'One has to cease believing that M.E. resembles or is the same as CFS. One has to cease believing those who patent virus and infectious agents for profit are necessarily going to tell the truth. If one looks only at M.E. and the diagnostic principles outlined above it is obvious that only one group of viruses can fit the picture as the causal agent of M.E. and those are the enterovirus family. There are no known patents on this group of viruses. The pathologies of acute onset M.E. patients as a group differ significantly from gradual onset patients CFS patients. Gradual onset CFS patients as a group have major and multiple missed organ and system pathologies rather than having typical M.E. Many CFS patients have no observable CNS findings.'
DR BYRON HYDE 2011

'Adrenaline surges are one of the best and worst things about M.E. They provide a way for our bodies to cope with overexertion in the short term and they can allow us to attend events that are very important to us (such as funerals, weddings and medical appointments) which we would normally be too ill for.

Unfortunately, they also let our bodies 'write cheques they can't cash' and are the reason why so many of us are severely affected. The payback for each adrenaline surge is just so enormous and so prolonged. It can be tempting to rely on them for a while especially when you are first ill, until the whole house of cards inevitably falls down and you are far more ill than when you started, possibly for months or years afterward. Adrenaline surges are also so often misunderstood by others.

For example, when I warn a friend that I am having a bad day and may not be up to much while we have our visit, the adrenaline surge phenomena of M.E. sometimes creates an illusion of good health. I get more and more ill as the evening wears on, and when it gets bad enough that my body is in real physiological difficulty, my body floods with adrenaline and I appear to suddenly become quite well. I talk a lot and very quickly. It is frustrating but of course quite understandable that so few people can see the difference between genuine health and vitality, and an adrenaline surge brought on by a health crisis, the latter of which is anything but a sign of good health in the M.E. patient. I don't expect others to always recognise this sign of a pending relapse, but to have it misinterpreted as a sign of improvement can be hard to take!'
JODI, M.E. PATIENT

'Dr Bruce Carruthers, who chaired the 2003 Canadian Clinical Case Definition for M.E./CFS, was also present when I gave this definition. I strongly disagreed with Dr Caruthers in the merging the definitions of M.E. and CFS since on the basis of the physical total body assessment of both M.E. and CFS patients; these two names represent two entirely different spectrums of illnesses. It is increasingly obvious that too much importance is being placed upon the definitions of Chronic Fatigue Syndrome (CFS), and not enough upon the actual disease, Myalgic Encephalomyelitis (M.E.). These two illness spectrums are not the same and should not be considered to be the same. Nor is there any doubt in my mind that the various definitions of CFS actively impede physicians' ability to make a rapid and rational diagnosis as well as a scientific confirmation of any testable illness. Such is not true of M.E. where a rapid and rational diagnosis can be made that can be confirmed by laboratory and other technological testing.'
DR BYRON HYDE 2011

CHAPTER TWO
More information on M.E.

This chapter includes the following two papers:

1. M.E. vs. M.S.: Similarities and differences – Condensed/modified version

Myalgic Encephalomyelitis (M.E.) and Multiple Sclerosis (M.S.) are very similar diseases medically in many ways. However, for reasons that have nothing to do with science, the two diseases are treated very differently politically and socially. The contrast could not be more stark.

M.E. patients are treated terribly (and often abused, even unto death in some cases), yet there is no public outcry as there would be if M.S. patients were treated in this same way. Thus people with M.E. find themselves in the position of actually envying people who have M.S.

2. What is M.E.? The full-length version of the paper.

A fully referenced historical, political and medical overview of M.E.

M.E. vs. M.S.: Similarities and differences - Condensed/modified version

 As many members of the public and the medical profession will be aware, Multiple Sclerosis (M.S.) is a disabling neurological disease which also affects the muscles. M.S. is a terrible disease and can cause severe disability and extreme suffering.

However, as surprising or bizarre as it seems, there is a section of the community which has reason to be envious of people who have M.S. It is made up of people who have the disabling neurological disease called Myalgic Encephalomyelitis (M.E.)

Medical similarities between M.S. and M.E.

M.E. and M.S. are actually very similar medically in many ways, as the following list demonstrates.

Table 1. Medical similarities between M.S. and M.E.

Multiple Sclerosis	Myalgic Encephalomyelitis
M.S. is primarily a neurological disease, i.e. a disease of the central nervous system (CNS).	M.E. is primarily a neurological disease, i.e. a disease of the central nervous system (CNS).
Demyelination (damage to the myelin sheath surrounding nerves) has been documented in M.S.	Demyelination (damage to the myelin sheath surrounding nerves) has been documented in M.E.
Evidence of oligoclonal bands in the cerebrospinal fluid has been documented in M.S.	Evidence of oligoclonal bands in the cerebrospinal fluid has been documented in M.E.
No single definitive laboratory test is yet available for M.S. but a series of tests are available which can objectively confirm the diagnosis with some certainty.	No single definitive laboratory test is yet available for M.E. but a series of tests are available which can objectively confirm the diagnosis with a high degree of certainty.
M.S. can be severely disabling and cause significant numbers of patients to be bedbound or wheelchair-reliant.	M.E. can be severely disabling and cause significant numbers of patients to be bedbound, wheelchair-reliant or housebound.
M.S. can be fatal (either from the disease itself or from complications arising from the disease).	M.E. can be fatal (either from the disease itself or from complications arising from the disease).
M.S. significantly reduces life expectancy.	M.E. significantly reduces life expectancy.
Symptoms/problems which occur in M.S. include: impaired vision, nystagmus, afferent pupillary defect, loss of balance and muscle coordination,	Symptoms/problems which occur in M.E. include: impaired vision, nystagmus, afferent pupillary defect, loss of balance and muscle coordination,

cogwheel movement of the legs, slurred speech, difficulty speaking (scanning speech and slow hesitant speech), difficulty writing, difficulty swallowing, proprioceptive dysfunction, abnormal sensations (numbness, pins and needles), shortness of breath, headaches, itching, rashes, hair loss, seizures, tremors, muscular twitching or fasciculation, abnormal gait, stiffness, subnormal temperature, sensitivities to common chemicals, sleeping disorders, facial pallor, bladder and bowel problems, difficulty walking, pain, tachycardia, stroke-like episodes, food intolerances and alcohol intolerance, and partial or complete paralysis.	cogwheel movement of the legs, slurred speech, difficulty speaking (scanning speech and slow hesitant speech), difficulty writing, difficulty swallowing, proprioceptive dysfunction, abnormal sensations (numbness, pins and needles), shortness of breath, headaches, itching, rashes, hair loss, seizures, tremors, muscular twitching or fasciculation, abnormal gait, stiffness, subnormal temperature, sensitivities to common chemicals, sleeping disorders, facial pallor, bladder and bowel problems, difficulty walking, pain, tachycardia, stroke-like episodes, food intolerances and alcohol intolerance, and partial or complete paralysis.
M.S. can cause orthostatic intolerance (dizziness or faintness on standing).	M.E. commonly causes severe orthostatic intolerance (which often worsens to become severe Postural Orthostatic Tachycardia Syndrome and/or Neurally Mediated Hypotension).
Short-term memory loss and other forms of cognitive impairment occur in 50% of M.S. patients. 10% of M.S. patients have cognitive impairments severe enough to significantly affect daily life.	Short-term memory loss and other forms of cognitive impairment occur in 100% of M.E. patients. Almost all M.E. patients have cognitive impairments that significantly affect daily life.
M.S. patients often become much more ill in even mildly warm weather. Cold weather can also cause significant problems.	M.E. patients often become much more ill in even mildly warm weather. Cold weather can also cause significant problems.
M.S. is thought to cause a breakdown of the blood brain barrier.	M.E. is thought to cause a breakdown of the blood brain barrier.
M.S. can affect autonomic nervous system function (including involuntary functions such as digestion and heart rhythms).	M.E. can affect autonomic nervous system function (including involuntary functions such as digestion and heart rhythms).
A positive Babinski's reflex is consistent with several neurological conditions, including M.S.	A positive Babinski's reflex (or extensor plantar reflex) is consistent with M.E.
The Romberg test will often be abnormal in M.S. (This test measures neurological dysfunction.)	The Romberg test will be abnormal in 95% or more of M.E. patients.
An abnormal neurological exam is usual in M.S. Abnormalities are also commonly seen in neuropsychological testing in M.S.	An abnormal neurological exam is usual in M.E. Abnormalities are also commonly seen in neuropsychological testing in M.E.
M.S. causes a certain type of brain lesion detectable in MRI brain scans. Abnormalities are also seen in EEG and QEEG brain maps and SPECT brain scans in M.S.	M.E. causes a certain type of brain lesion detectable in MRI brain scans. Abnormalities are also seen in EEG and QEEG brain maps and SPECT brain scans in M.E.
Hypothyroidism is found in many M.S. patients.	Hypothyroidism is found in almost all M.E. patients.
The glucose tolerance test is often abnormal in M.S.	The glucose tolerance test is often abnormal in M.E.

Low blood pressure readings (usually low-normal) are common in M.S.	Low blood pressure readings are extremely common in M.E. Severely low blood pressure readings as low as, or lower than, 84/48 are common in severe M.E. or those having severe relapses. This can occur at rest or as a result of orthostatic or physical overexertion. Circulating blood volume measurements of only 50% to 75% of expected are also commonly seen in M.E.
Patients with M.S. have an increased risk of dying from heart disease or vascular diseases.	Deaths from cardiac problems are one of the most common causes of death in M.E.
Although M.S. is primarily neurological, it also has aspects of autoimmune disease.	Although M.E. is primarily neurological, it also has aspects of autoimmune disease.
M.S. usually affects people between the ages of 20 and 40 years, and the average age of onset is approximately 34 years. Onset occurs between the ages of 20 to 40 years in 70% of patients.	The average ages affected by M.E. are similar to those seen in M.S. However, the average age of *onset* may be significantly younger in M.E.
M.S. was once thought to be rare in children, but we now know that around 5% of M.S. sufferers are under 18.	Around 10% of M.E. sufferers are under 18.
M.S. affects more than a million adults and children worldwide.	M.E. affects more than a million adults and children worldwide.

As well as there being many similarities in symptoms, the brain scans from M.E. and M.S. patients are often very similar. M.S. and M.E. both cause a certain type of brain lesion detectable in brain scans. Those with M.S. tend to have fewer brain lesions of a larger size, while M.E. is associated with a greater number of these lesions of a somewhat smaller size.

M.E. and M.S. are so similar medically that they are sometimes misdiagnosed as one another.

The names used for M.E. and M.S. also indicate the similarities between the two diseases. M.S. was first described in 1868, and M.S. has also been known as 'disseminated sclerosis' or 'encephalomyelitis disseminate.' Myalgic Encephalomyelitis has existed for centuries but was first comprehensively scientifically documented in 1934, when an outbreak of what at first seemed to be Poliomyelitis (Polio) occurred in Los Angeles (M.E. occurs in outbreaks as well as sporadically).

The term Myalgic Encephalomyelitis was coined in 1956. Earlier names for M.E. include 'atypical Polio' and atypical Multiple Sclerosis.'

Both M.S. and M.E. have been correctly classified as organic diseases of the central nervous system in the World Health Organization's International Classification of Diseases for many decades. M.S. is classified at G 35 and M.E. at G 93.3.

Why are people with M.E. often envious of people with M.S.?

M.S. and M.E. are distinct diseases, but they are in many ways very similar medically.

However, despite the medical similarities, the two diseases are treated very differently politically and socially.

The differences between the political and social treatment of M.S. and M.E. are the reason for M.E. patients' envy of M.S. sufferers.

Table 2. Political and social differences between the treatment of M.S. and M.E.

Multiple Sclerosis	Myalgic Encephalomyelitis
M.S. is a neurological disease, so M.S. patients are treated primarily by neurologists. In countries such as Australia, Canada, New Zealand, the USA and the UK, the majority of M.S. patients have access to a neurologist who is knowledgeable about M.S.	M.E. is also a neurological disease that is appropriately treated by a neurologist, yet very few M.E. patients have access to a doctor who knows even the most basic facts of M.E., let alone access to a neurologist who has experience and knowledge of M.E. The vast majority of M.E. patients have no access to appropriate medical care at all.
Media reports on M.S. are of a high standard. If reporters put out stories about M.S. that were not factual, there would be a public outcry and then an apology made.	Media reports on M.E. are of a very low standard. It is extremely common to read articles claiming to be about M.E. but which do not contain even one accurate fact about the disease. Complaints made by M.E. patients and experts are ignored.
Media reports on those who have experienced some recovery from M.S. involve genuine M.S. patients.	Media reports of 'miracle recoveries' from M.E. touting one pseudo-treatment or another are very common. However, the patients described did not have M.E. (or any other serious neurological disease).
M.S. advocacy groups do good work for M.S. patients and help raise awareness and funds for research. M.S. groups are run for and by M.S. patients.	The vast majority of M.E. advocacy groups do not advocate on behalf of M.E. patients but instead work directly against the best interests of M.E. patients. The vast majority of these groups are **not** run for or by M.E. patients and their agendas are **not** helping M.E. patients. These groups often distribute information on M.E. which is completely inaccurate and which also belittles and misrepresents M.E. patients.
M.S. charities would never support treatments for M.S. which had zero chance of success, and which very often caused a severe and prolonged deterioration of the disease, or even death.	So-called M.E. charities very often fully support and push 'treatments' for M.E. which have zero chance of success, and which very often cause a severe and prolonged deterioration of the disease, or even death.
M.S. is a well-known illness, and patients are generally treated appropriately by doctors and other medical staff.	M.E. is an illness that most medical staff are not well educated about. M.E. patients are often treated inappropriately by doctors and other medical staff.
People with M.S. will generally qualify for the welfare and medical insurance payouts they are entitled to.	People with M.E. are often denied the appropriate welfare and medical insurance payouts they are entitled to.
M.S. receives many millions of dollars in government funding for research, and millions more are raised each year by M.S. charities around the world.	M.E. receives no government funding worldwide and very little is raised by charities. What little is raised by these groups is virtually always spent researching non-M.E. patient groups or mixed patient-groups, but even those studies which do include a small proportion of M.E. patients are useless, because mixed patient groups make any results meaningless.

When research says it involves M.S. patients one can have a high degree of confidence that this is indeed the case.	When research says it involves M.E. patients one can only have a low degree of confidence that this is indeed the case.
When a patient with M.S. chooses euthanasia, public sympathy is expressed for the degree of pain and suffering that must have led to such a choice.	When a patient with M.E. chooses euthanasia, public derision is often expressed. Even when, as is almost always the case with euthanasia, the person with M.E. was severely affected and bedbound, it is very often blithely claimed by the media either that the patient was not ill at all, or had a mild disease that could be easily cured within weeks *if the patient only truly wanted to get better.*

Although M.S. and M.E. are very similar medically, they are worlds apart politically and socially.

Why is the public perception of M.E. so different to that of M.S.?

The public perception of M.S. reflects the reality. Most members of the public are aware of the basic facts: that M.S. is a neurological disease affecting the muscles, and that it can be very disabling or fatal. Understanding of these facts is also reflected in the way the media handles M.S. and government policy on M.S.

The public perception of M.E. could not be further removed from the medical reality of the disease.

Most members of the public, if they have heard of M.E., have heard an entirely inaccurate account of the disease which they mistakenly believe to be based on science. Despite the fact that M.E. is a serious neurological disease comparable to M.S., Lupus and Polio, M.E. is seen by most of the public and even by most of the medical profession as 'trivial.' M.E. is perceived and presented similarly by most of the media and by government. M.E. patients are treated differently to those with comparable diseases such as M.S. The contrast is stark.

There is an abundance of evidence showing that M.S. is an organic neurological disease that can be severely disabling or fatal. The same is true of M.E. The evidence supporting M.E. is no less compelling, although you would not know this from the way M.E. is dealt with. If anything, M.E. has more scientific credibility; it is far easier to diagnose due to its acute onset and more obvious, systemic and unique pathology; the cause is far more certain in M.E.

In short, the reason M.E. is treated so differently to M.S., despite their being comparable diseases, has nothing to do with science or evidence, and everything to do with MONEY.

M.E. patients are being (mis)treated based purely on financial considerations. Financial vested interest groups have subverted and obscured the reality of M.E. for their own benefit. Many millions of dollars are being made (or saved) by powerful medical insurance companies, and others, by this scam. (This is explained in detail in <u>What is M.E.?</u>)

This lucrative fiction about M.E. is widely accepted in the community, but it has about as much to do with science as astrology has with astronomy. It has been scientifically disproven hundreds of times over, even *before* the large scale cover-up/scam was created.

Repeating a lie over and over again will never make it true, but it seems it often will make lots of people believe it to be true, especially if the sources are seen as 'authorities.' **That's why this abusive and unscientific money-making fiction about M.E. has continued for 20 years and has only become more extreme and entrenched over time.**

M.S. is not being targeted in the same way as M.E. by insurance companies etc. This is a matter of timing: M.S. emerged earlier, received more medical attention and has been longer established within mainstream medicine than M.E. Because of this fortunate timing difference, M.S. has escaped the modern manipulation for profit which has plagued M.E. for the last 20 years.

For example, doctors working for medical insurance companies are able to get influential government advisory positions in the field of health which play a large role in determining how diseases are treated, categorized and defined. Giving corporations with vested interests the power to unscientifically re-define and/or re-classify (i.e. wrongly re-classify) a disease to suit their own interests can be *immensely* lucrative for them. Political interests have determined how M.E. is dealt with and how it is perceived, which is not true of M.S.

The reason M.E. patients are so poorly and inappropriately treated is clear. How to stop this abuse, when governments, so-called M.E. charities and the media are colluding in a cover-up for their own benefit, is far less clear.

How patients with M.E. can work to change the situation when they are so ill and disabled, and when so many are too ill to even be able to read the facts about what is happening, and when they have so little other support, is not clear. How can patients with M.E. get through to the vast majority of the public who refuses to believe government and industry could be so immoral (despite ample examples of past transgressions)? How can patients with M.E. convince others of the truth when so many seemingly benign companies, government departments, journalists or supposedly patient-based organisations are producing so much mutually supportive and superficially convincing propaganda?

These are hard questions and simply enormous problems which M.E. patients are forced to deal with, and which M.S. patients need not ever consider.

M.E. patients sometimes wonder how their lives would be different if they instead had M.S....

People with M.E. wonder if those with M.S. know how lucky they are to be able to go to the emergency room when they are very unwell, and in fear of dying, and know that they will be treated with respect and given the appropriate care; rather than laughed at, mocked in front of the other patients, refused tests or treatment and just sent home. M.E. sufferers wonder if M.S. patients know how lucky they are that millions of dollars are being spent trying to cure their disease. That knowledge must be so comforting.

People with M.E. wonder if those with M.S. know how lucky they are to have access to a doctor who knows at least the basic facts of their disease. Very few patients with M.E. have such a 'luxury.' Some M.E. patients wonder if M.S. patients know how lucky they are to not have to worry that the latest unscientific study or article that claims to be about their disease will cause those around them to mistreat them due to the study involving an unrelated patient group, which suits the authors own vested interests.

M.S. can be very severe. So can M.E. Being severely ill is hard, but being severely ill through mistreatment, apathy and neglect – as is the case with many severe M.E. patients – is even harder to deal with.

Severe injury (or death) is inflicted on thousands of people with M.E. every year because of inappropriate medical advice, but this does not seem to cause any public concern. There would be outrage if even a tiny fraction of the harm done to M.E. patients was done to people with other diseases, but the outrage is just not there for us. For M.E. patients *that* is very, very hard to live with.

It is also hard for M.E. patients to live with the fact that some people with M.E. are reduced to poverty by refusal of welfare or insurance payments, which they would have received if they had M.S. That some people with M.E. have died from inappropriate and cruel medical mistreatment, and their abusers will never be brought to justice. That some M.E. patients have had their family and friends disown them due to misconceptions about M.E. That some M.E. patients have their rights taken away, and subjected to treatments that cannot improve their condition but which carry an enormous risk of worsening the disease seriously, or causing death. That some parents of children with M.E. have been charged with causing their child's illness (falsely accused of Munchausen by Proxy) and had their children removed from their care and then seriously medically abused. That small children very ill with M.E. have been thrown into swimming pools (and very nearly drowned), or denied food or contact with family, in an attempt to force them to do things they are too ill to do because of their disease.

M.E. is one of the most severely disabling and devastating diseases there is. Yet despite all the medical advances in today's high-tech world, it is as though M.E. patients live in another era and receive only the most primitive and rudimentary care – if indeed they receive any care at all.

Conclusion

People will often say to M.E. patients 'at least you don't have M.S. It could be worse, you should be grateful.' But if anything the opposite is true. Taking everything into account, the physical reality of each disease plus the misconceptions and mistreatment associated with M.E., it is very hard to see how *anyone* would ever choose M.E. over M.S., if such a choice were possible.

That isn't to say that M.S. isn't a terrible disease, or that those with M.S. have any more resources or funding than they rightly deserve and need, or that everyone with M.S. always gets every service they need easily and will always have a very supportive family. *Of course not.* The point is that it makes no sense that patients with these two very similar diseases are treated so differently just because of political manipulation for profit; that scientific reality, ethics and logic count for so little.

It is as bizarre and unfair as if those with broken arms were given x-rays and had the broken bone set and put in a cast until it had healed, while those with broken legs were told to go home and stop wasting the doctor's time, or that perhaps taking up jogging would make them feel better.

M.S. is a disabling neurological disease that causes a high degree of suffering. So is M.E. However, there is a whole other world of suffering experienced by M.E. patients which is unknown to M.S. patients and others with diseases where the public perception and political treatment of the disease is closely aligned with the medical reality. It is an additional type of suffering which can be as much a burden as the disease itself. When you combine these political problems with a disease as serious as M.E., it makes M.E. hell on earth.

The treatment of people with M.E. must be based on science at last, as is the treatment of M.S. patients. All M.E. patients want is to be treated the same way as those with M.S. and other comparable illnesses. All M.E. patients want is for studies on M.E. to actually involve M.E. patients, for the term M.E. to only be used to describe actual M.E. patients, for the facts about M.E. to be taught at medical schools in the same way M.S. facts are, for appropriate money to be made available for M.E. research, for government policy on M.E. to reflect the reality of M.E., and for the media (including medical journals) to write articles about M.E. with the same standard of factual accuracy as articles on M.S., and other diseases.

These things don't seem much to ask for in this day and age, but right now, they out of reach for M.E. patients and if anything they get further and further away as each year passes.

More information and additional notes on this text

- For more information on this topic, including a table describing the medical *differences* between M.S. and M.E., please see the full-length version of this text on the HFME website. Please also note that none of these charts is designed to be comprehensive or detailed enough to be used to differentiate between an M.S. or M.E. diagnosis. Additional resources on Multiple Sclerosis used in creating this paper are listed in the full-length version of this text on the website.

- *An important note:* To be clear, while M.E. can be far more disabling than M.S. and some medical aspects of it are worse than is seen in M.S., there is no doubt that *some aspects* of M.S. are *undoubtedly worse* than *some aspects* of M.E. The level of suffering can be very high in both diseases, and nobody with either disease has it easy. Those with the worst deal illness-wise are those with the most severe forms of either M.E. or M.S.

    To read this paper and feel that the statement being made is that M.E. is always worse than M.S., and that M.S. does not cause immense suffering or doesn't deserve more funding for support, is to miss the point of this paper entirely.

    People with M.E. and M.S. have been dealt a very cruel blow and need and deserve all the support and kindness they can get. Both are absolutely devastating diseases.

To read a fully-referenced version of the medical information in this text compiled using information from the world's leading M.E. experts, please see the 'What is M.E.?' paper in this book or on the HFME website.

Acknowledgments

Thanks to Lesley Ben for editing this paper.

Relevant quotes

'I have a friend who has M.S. and she is really very independent and able to get about (unlike me). My Dr isn't very helpful but at least she is kind. My friend with M.S. has said how similar our illnesses are, yet she has a special M.S. nurse, a supportive Dr and the knowledge that when she tells someone she has M.S. they will be understanding and non-judgemental.

   If only we could have half of that I would be happy. I had to use a wheelchair for 2 years and I had people saying I was lazy and why was I 'carrying on' like that! Can you believe it?! When I was so severely ill and couldn't get out of bed for months on end people told my husband I needed 'motivating' and that I was probably having a nervous breakdown and was depressed... If only I had M.S.!!!

   I really do feel that we are left to cope with such a debiltating illness alone almost. My husband has nearly lost his job because of time off he has had to take to care for me when I'm at my worst. We struggle financially as we only have one wage coming in and yet we get no help or support from anywhere, and yet if I was suffering from M.S. people would be appalled at my situation. As it is, most people don't even think I'm ill and that I should just 'pull myself together' - if only I could! And this is the joke, we are too _____ ill to stand up for our rights and make a change!'
LENA, M.E. PATIENT

'i have to admit i get very resentful when i see the adverts on television for everything except M.E. (not to deny serious diseases their rightful place in public awareness, but it's hard to be left out in the cold year after year, decade after decade). on a bad day i probably feel as wretched as people with severe M.S.; on a "good" day i probably feel about as well as people with a mild case of M.S. but then it's probably the same for them ... degrees of severity.

   having struggled with M.E. in the face of doubt, invalidation, "pep talks", psychologizing, and so on since i was 28 ... i'll be turning 55 next month ... i think that had someone come to me and given me the choice, i'd have chosen M.S. for exactly the reasons put forth by so many others. that M.E. isn't recognised as "real" (despite the mountains of evidence to the contrary) often pushes me to the brink of absolute despair, and seeing so many other people recceiving the kind of care i need just adds insult to injury.

   i have long since lost count of the times i've been scolded and sent home untreated, only to return a few days later with symptoms no one can deny or ignore. I fully expect my cause of death will be medical neglect.'
NAMASTE, SEVERE M.E. PATIENT

'I do think that the simultaneous rise of AIDS at the same time as the 80's US M.E. epidemics sucked the life out of any possible attention by public health towards M.E. It is so hard to talk about because AIDS is a TERRIBLE disease but they got the funding and the recognition and research because it was a very powerful, large, population group here in NA, well used to activism, and by the way, men, at least in the early days. And just look what has happened for PWA's, all the research and progress, all the social support and awareness. There's quite a strong AIDS group here. I found out that a woman I know with it, who has a husband and kids, who functions as healthy running around all over the place, and who gets substantial caregiver hours. And here I am, unable to do my own shopping or housework and can't get one lousy hour. And one can never talk about it, it's not PC <sigh>. I do not begrudge them their care etc. but only wish for 1/10th of the help that they get (or that those with M.S. get).'
AYLWIN CATCHPOLE, M.E. PATIENT 20+ YEARS (CANADA)

'I've just read your M.E. vs M.S. page. 20 yrs ago I was treated and tested by a neurologist who thought I had M.S. When I was diagnosed with M.E. instead of M.S., I was thought I was fortunate, but often since then I've wished the M.S. diagnosis was the correct one instead of M.E.'
WENDY, M.E. PATIENT

'I actually walked away from mainstream medical care about 15 yrs ago out of utter frustration. Now I am having another "go" just to try and obtain any help that might be had. And my perception is that it's worse now. And my patience is running out. again. I almost wish I could be MISdiagnosed with M.S. or other, more respected disease, just to get better treatment. No offense to anybody, I have known people with M.S.,

Lupus, Lyme, AIDS (not just asymptomatic HIV) and all kinds of other similar conditions, and to a one they can all run circles around me...until it's time to die that is. (Then they are more disabled than us, but only then). <sigh>'
AYLWIN CATCHPOLE, M.E. PATIENT 20+ YEARS (CANADA)

'I have had M.S. for 10 years and I have been amazed by the similarities to people I know with M.E. and I fully sympathise with the issues regarding the different attitude of the public and professionals to the two illnesses and about the different levels of support that are available. Thank you for the article and I wish everyone with M.S. or M.E. the courage to deal with each day.'
PAM, M.S. PATIENT

'I don't think there actually are any other disease-sufferers getting as little help as we do. And yet there are very few diseases (if any), which impact, destroy and restrict your life and ability to function as much as M.E. does. I would not get any caregiver hours or assistive devices either, if it was not for the fact that I also have a connective tissue disorder called the Ehlers-Danlos syndrome (EDS). EDS is not the actual reason for my disability and need of help. The reason for me needing help and being so sick and disable is M.E. But with "only" M.E. I would not get ANY kind of help. Regardless of the fact that I would not stay alive without any caregiver hours.'
M, SEVERE M.E. PATIENT

"It has become obvious to me that we are dealing with both a vasculitis and a change in vascular physiology. Numerous other physicians have supported this finding. M.E. appears to be in this same family of diseases as paralytic polio and M.S. M.E. is less fulminant than M.S. but more generalized. M.E. is less fulminant but more generalized than poliomyelitis. This relationship of M.E.-like illness to poliomyelitis is not new and is of course the reason that Alexander Gilliam, in his analysis of the Los Angeles County General Hospital M.E. epidemic in 1934, called M.E. atypical poliomyelitis.'
DR BYRON HYDE 2006

'There is ample evidence that M.E. is primarily a neurological illness. It is classified as such under the WHO international classification of diseases (ICD 10, 1992) although non neurological complications affecting the liver, cardiac and skeletal muscle, endocrine and lymphoid tissues are also recognised. Apart from secondary infection, the commonest causes of relapse in this illness are physical or mental over exertion.'
DR ELIZABETH DOWSETT

'Possible costing for M.E. support has been based on 3 times the cost of maintenance for Multiple Sclerosis on the supposition that M.E. is [up to] 3 times as common. The only costs that we can be sure of are those derived from the failure of appropriate management, and of inappropriate assessments which waste vast sums of money and medical time while allowing patients to deteriorate unnecessarily. Research workers must be encouraged and appropriately funded to work in this field. However they should first be directed to papers published before 1988, the time at which all specialised experience about poliomyelitis and associated infections seem to have vanished mysteriously!'
DR ELIZABETH DOWSETT

'People in positions of power are misusing that power against sick people and are using it to further their own vested interests. No-one in authority is listening, at least not until they themselves or their own family join the ranks of the persecuted, when they too come up against a wall of utter indifference.'
PROFESSOR M. HOOPER 2003

There are many clinical and laboratory similarities in M.E. and M.S., but what separates them is: the plethora of systemic manifestations in M.E., the orthostatic tachycardia seen in M.E., the outbreaks of M.E., the striking involvement of muscle in M.E. and the muscle pathology seen in M.E., the characteristic myalgias and arthralgias in M.E., and the symptoms such as cold extremities and flu-like symptoms etc. seen in M.E. These features are not seen in M.S. and their presence may even preclude a M.S. diagnosis.
FROM CHARLES M POSER MD IN THE BOOK THE CLINICAL AND SCIENTIFIC BASIS OF MYALGIC ENCEPHALOMYELITIS (PARAPHRASED BY THE AUTHOR)

What is M.E.? A historical, medical and political overview
COPYRIGHT © JODI BASSETT 2004. UPDATED DECEMBER 2011. FROM WWW.HFME.ORG

 Myalgic Encephalomyelitis (M.E.) is a debilitating acquired neurological disease which has been recognised by the World Health Organisation (WHO) since 1969 as a distinct organic neurological disorder.

M.E. can occur in both epidemic and sporadic forms, and over 60 outbreaks of M.E. have been recorded worldwide since 1934. M.E. is similar in a number of significant ways to Multiple Sclerosis, Lupus and Poliomyelitis (Polio).

M.E. can be extremely severe and disabling and in some cases the disease is fatal.

Is M.E. a new illness? What does the name Myalgic Encephalomyelitis mean?

The disease we now know as Myalgic Encephalomyelitis is not a new disease. M.E. is thought to have existed for centuries (Hyde 1998, [Online]) (Dowsett 1999a, [Online]).

In 1956 the name Myalgic Encephalomyelitis was created. The term was invented jointly by Dr A Melvin Ramsay, who coined this name in relation to the Royal Free Hospital epidemics that occurred in London in 1955 – 1957, and by Dr John Richardson, who observed the same type of illness in his rural practice in Newcastle-upon-Tyne during the same period. It was obvious to these physicians that they were dealing with the consequences of an epidemic and endemic infectious neurological disease (Hyde 1998, [Online]) (Hyde 2006, [Online]).

The term Myalgic Encephalomyelitis means: My = muscle, algic = pain, encephalo = brain, mye = spinal cord, itis = inflammation (Hyde 2006, [Online]).

As M.E. expert Dr Byron Hyde writes:

> The reason why these physicians were so sure that they were dealing with an inflammatory illness of the brain is that they examined patients in both epidemic and endemic situations with this curious diffuse brain injury. In the epidemic situation with patients falling acutely ill and in some cases dying, autopsies were performed and the diffuse inflammatory brain changes are on record (2006, [Online]).

The Wallis description of M.E. was created in 1957, and in 1959 Sir Donald Acheson (a former UK Chief Medical Officer) conducted a major review of M.E.

In 1962 the distinguished neurologist Lord Brain included M.E. in the standard textbook of neurology. In recognition of the large body of compelling research that was available, M.E. was formally classified as an organic disease of the central nervous system in the World Health Organisation's International Classification of Diseases in 1969.

In 1978 the Royal Society of Medicine held a symposium on Myalgic Encephalomyelitis at which M.E. was accepted as a distinct entity. The symposium proceedings were published in The Postgraduate Medical Journal later that same year. The Ramsay case description of M.E. was published in 1981 (Hooper et al. 2001, [Online]).

Since 1956 the term Myalgic Encephalomyelitis has been used to describe the illness in the UK, Europe Canada and Australasia. This term has stood the test of time for more than 50 years. The recorded medical history of M.E. as a debilitating organic neurological illness affecting children and adults is substantial; it spans over 80 years and has been published in prestigious peer-reviewed journals all over the world (Hyde 1998, [Online]) (Hooper 2003a, [Online]) (Dowsett 2001b, [Online]).

As award winning microbiologist and M.E. expert Dr Elizabeth Dowsett explains: 'There is ample evidence that M.E. is primarily a neurological illness, although non-neurological complications affecting the liver, cardiac and skeletal muscle, endocrine and lymphoid tissues are also recognised' (n.d.b, [Online]).

M.E. is not defined by mere 'fatigue'

M.E. is not synonymous with being tired all the time. If a person is very fatigued for an extended period of time this does not mean they are having a 'bout' of M.E. Such a suggestion is no less absurd than to say that prolonged fatigue means a person is having a 'bout' of Multiple Sclerosis, Parkinson's disease or Lupus. If a person is constantly fatigued this should not be taken to mean that they have M.E., no matter how severe or prolonged their fatigue is.

Fatigue is a symptom of many different illnesses as well as a feature of normal everyday life – but it is not a defining symptom of M.E., or even an essential symptom of M.E. The terms 'fatigue' and 'chronic fatigue' were not associated with defining this illness until the entity of 'Chronic Fatigue Syndrome' was created in 1988 in the USA (Hyde 2006, [online]). But M.E. and 'CFS' are *not* synonymous terms.

'Fatigue' and 'feeling tired all the time' are not at all the same thing as the very specific type of *paralytic muscle weakness* or *muscle fatigue* which *is* characteristic of M.E. (caused by mitochondrial dysfunction) and which affects every organ and cell in the body, including the brain and the heart. This causes – or significantly contributes to – such problems in M.E. as cardiac insufficiency (a type of heart failure), orthostatic intolerance or POTS (inability to maintain an upright posture), blackouts, reduced circulating blood volume (and pooling of the blood in the extremities), seizures (and other neurological phenomena), memory loss, problems chewing/swallowing, episodes of partial or total paralysis, muscle spasms/twitching, extreme pain, problems with digestion, vision disturbances, and breathing difficulties.

These problems are exacerbated by even trivial levels of physical and cognitive activity, sensory input and orthostatic stress beyond a patient's individual limits. People with M.E. are made very ill and disabled by this problem with their cells; it affects virtually every bodily system and has also lead to death in some cases. Many patients are housebound and bedbound and are often so ill that they feel they are about to die. People with M.E. would give *anything* to only be severely 'fatigued' or 'tired all the time' (Bassett 2010, [Online]).

Fatigue, post-exertional fatigue or malaise may occur in many different illnesses such as various post-viral fatigue states or syndromes, Fibromyalgia, Lyme disease, and many others, but what is happening with M.E. patients is an entirely different and unique problem of a much greater magnitude. These terms are not accurate or specific enough to describe what is happening in M.E.

M.E. is a neurological illness of extraordinarily incapacitating dimensions that affects virtually every bodily system – not a problem of 'chronic fatigue' (Hyde 2006, [Online]) (Hooper 2006, [Online]) (Hooper & Marshall 2005a, [Online]) (Hyde 2003, [Online]) (Dowsett 2001, [Online]) (Hooper et al. 2001, [Online]) (Dowsett 2000, [Online]) (Dowsett 1999a, 1999b, [Online]) (Dowsett 1996, p. 167) (Dowsett et al. 1990, pp. 285-291) (Dowsett n.d., [Online]).

- For more information see M.E. is not fatigue, or 'CFS'. Many of the world's leading M.E. experts have spoken out strongly against claims that 'fatigue' is the defining/essential symptom of M.E. See M.E. is not defined by 'fatigue' to read some of their comments. For more information on the symptoms of M.E., including the unique reaction people with M.E. have to activity, see: The ultra-comprehensive M.E. symptom list.

If M.E. and 'CFS' are not synonymous terms, why do some groups claim that they are? What is 'CFS'?

The disease category of 'CFS' was created in a response to an outbreak of what was unmistakably M.E., but this new name and definition did not describe the known signs, symptoms, history and pathology of M.E. It described a disease process that did not, and could not exist.

Why was the renaming and redefining of the distinct neurological disease M.E. allowed to become so muddied? Indeed, why did Myalgic Encephalomyelitis suddenly need to be renamed or redefined at all?

The answer is money. There was an enormous rise in the reported incidence of M.E. in the late 1970s and 1980s, alarming medical insurance companies in the US. So it was at this time that certain psychiatrists and others involved in the medical insurance industry (on both sides of the Atlantic) began their campaign to reclassify M.E. as a psychological or 'personality' disorder, in order to side-step the financial responsibility of so many new claims (Marshall & Williams 2005a, [Online]).

As Professor Malcolm Hooper explains:

> In the 1980s in the US (where there is no NHS and most of the costs of health care are borne by insurance companies), the incidence of M.E. escalated rapidly, so a political decision was taken to rename M.E. as "chronic fatigue syndrome", the cardinal feature of which was to be chronic or ongoing "fatigue", a symptom so universal that any insurance claim based on "tiredness" could be expediently denied. The new case definition bore little relation to M.E.: objections were raised by experienced international clinicians and medical scientists, but all objections were ignored... To the serious disadvantage of patients, these psychiatrists have propagated untruths and falsehoods about the disorder to the medical, legal, insurance and media communities, as well as to government Ministers and to Members of Parliament, resulting in the withdrawal and erosion of both social and financial support [for M.E. patients]. Influenced by these psychiatrists, government bodies around the world have continued to propagate the same falsehoods with the result that patients are left without any hope of understanding or of health service provision or delivery. As a consequence, government funding into the biomedical aspects of the disorder is non-existent (2003a, [Online]) (2001, [Online]).

The psychiatrist Simon Wessely – arguably the most powerful and prolific author of papers which claim that M.E. is merely a psychological problem of 'fatigue' – began his rise to prominence in the UK at the same time the first CFS definition was being created in the USA (1988). Wessely, and his like-minded colleagues – a small group made up mostly but not exclusively of psychiatrists (colloquially known as the 'Wessely School') has gained dominance in the field of M.E. in the UK (and increasingly around the world) by producing vast numbers of papers which purport to be about M.E.

Wessely claims to specialise in M.E. but uses the term interchangeably with chronic fatigue, fatigue or tiredness, plus terms such as neurasthenia, CFS and 'CFS/ME' (a confusing and misleading term he created himself). He claims that psychiatric states of ongoing fatigue and the distinct neurological disorder M.E. are synonymous. Despite all the existing contradictory evidence, Wessely (and members of the Wessely School) assert that M.E. is a behavioural disorder, with no physical signs of illness or abnormalities on testing, that is perpetuated by 'aberrant illness beliefs' 'the misattribution of normal bodily sensations,' and that patients 'seek and obtain secondary gain by adopting the sick role' (Hooper & Marshall 2005a, [Online]).

The Wessely School and collaborators have assiduously attempted to obliterate recorded medical history of M.E. even though the existing evidence and studies were published in prestigious peer-reviewed journals and span over 70 years. Wessely's claims, and those of his colleagues around the world, have flooded the worldwide literature to the extent that medical journals rarely contain any factual and unbiased information about M.E. Most clinicians are effectively being deprived of the opportunity to obtain even the most basic facts about the illness.

For at least a decade, serious questions have been raised in international medical journals about possible scientific misconduct and flawed methodology in the work of Wessely and his colleagues. It is only relatively recently however that his long-term involvement as medical adviser – and board member – to a number of commercial bodies with a vested interest in how M.E. is managed have been exposed.

This is the sole reason the myth that M.E. is a psychiatric or behavioural 'fatiguing' disorder or even an 'aberrant belief system' continues: not because there is good scientific evidence for the theory, or because the evidence proving organic causes and effects is lacking, but because such a theory is so **financially and politically convenient and profitable** on such a large scale to a number of extremely powerful corporations (Hooper et al 2001, [Online]).

As Dr Elizabeth Dowsett observes, these financially motivated theories bear as much relation to legitimate science as astrology does to astronomy (1999b [Online]). Professor Malcolm Hooper goes on to explain:

> Increasingly, it is now "policy-makers" and Government advisers, not experienced clinicians, who determine how a disorder is classified and managed in the NHS: the determination of an illness classification and the provision of policy-driven "management" is a very profitable business. To the detriment of the sick, the

deciding factor governing policies on medical research and on the management and treatment of patients is increasingly determined not by medical need but by economic considerations. There is a gross mismatch between the severity and complexity of M.E. and the medical and public perception of the disorder (2003a, [Online]).

Members of the 'Wessely school' in the UK including Wessely, Sharpe, Cleare and White, their US counterparts Reeves, Straus etc of the CDC, in Australia Lloyd, Hickie etc and the clinicians of the Nijmegen group in the Netherlands each support a bogus psychiatric or behavioural paradigm of 'CFS' and recommend rehabilitation-based approaches such as cognitive behavioural therapy (CBT) and graded exercise therapy (GET) as the most useful interventions for 'CFS' patients.

It is important to be aware that none of these groups is studying patients with M.E. Each of these groups uses a definition of 'CFS,' or has created their own, which does not select those with M.E. but instead selects those with various types of psychiatric and non-psychiatric fatigue. These inappropriate interventions are at best useless and at worst extremely harmful or fatal for M.E. patients.

The creation of the bogus disease category 'CFS' has been used to impose a false psychiatric paradigm of M.E. by allying it with various unrelated psychiatric fatigue states and post-viral fatigue syndromes for the benefit of various (proven) financial and political interests. The resulting 'confusion' between the distinct neurological disease M.E. and the bogus disease category of 'CFS' has caused an overwhelming additional burden of suffering for those who suffer from M.E. and their families.

It's a huge mess, that is for certain — but it is not an *accidental* mess — that is for certain too (Hyde 2006a, [Online]) (Hooper 2006, [Online]) (Hyde 2003, [Online]) (Hooper 2003a, [Online]) (Dowsett 2001a, [Online]) (Hooper et al. 2001, [Online]) (Dowsett 2000, [Online]) (Dowsett 1999a, 1999b, [Online]).

- To read about the vast difference between M.E. and 'CFS' (and how such a small (but powerful) group of vested interest psychiatrists have come to influence the opinions of the worldwide medical community about M.E.) see: Who benefits from 'CFS' and 'ME/CFS'? and Smoke and mirrors on HFME and also A Brief History of Myalgic Encephalomyelitis & An Irreverent History of CFS by Dr Byron Hyde
- For information on how the 'CFS' scam affects all parts of an M.E. patient's life, see M.E.: The shocking disease.

What does a diagnosis of 'CFS' actually mean?

There are now more than nine different definitions of 'CFS.' Each of these flawed 'CFS' definitions 'define' a heterogeneous (mixed) population of people with various misdiagnosed psychiatric and non-psychiatric states which have little in common but the symptom of fatigue.

The fact that a person qualifies for a diagnosis of 'CFS', based on any of the 'CFS' definitions: (a) does not mean that the patient has M.E., and (b) does not mean that the patient has any other distinct and specific illness named 'CFS.'

A diagnosis of' CFS' – based on any of the 'CFS' definitions – can only ever be a MISdiagnosis. All a diagnosis of 'CFS' actually means is that the patient has a gradual onset fatigue syndrome which is usually due to a *missed major disease*. As Dr Byron Hyde explains, the patient has:

a. Missed cardiac disease, b. Missed malignancy, c. Missed vascular disease, d. Missed brain lesion either of a vascular or space occupying lesion, e. Missed test positive rheumatologic disease, f. Missed test negative rheumatologic disease, g. Missed endocrine disease, h. Missed physiological disease, i. Missed genetic disease, j. Missed chronic infectious disease, k. Missed pharmacological or immunization induced disease, l. Missed social disease, m. Missed drug use disease or habituation, n. Missed dietary dysfunction diseases, o. Missed psychiatric disease (2006, [Online]).

Under the cover of 'CFS' certain vested interest groups have assiduously attempted to obliterate recorded medical history of M.E., even though the existing evidence has been published in prestigious peer-reviewed journals around the world and spans over 70 years. Dr Byron Hyde explains:

Do not for one minute believe that CFS is simply another name for Myalgic Encephalomyelitis. It is not. The CDC 1988 definition of CFS describes a non-existing chimera based upon inexperienced individuals who lack any historical knowledge of this disease process. The CDC definition is not a disease process. It is (a) a partial mix of infectious mononucleosis /glandular fever, (b) a mix of some of the least important aspects of M.E. and (c) what amounts to a possibly unintended psychiatric slant to an epidemic and endemic disease process of major importance. Any disease process that has major criteria, of excluding all other disease processes, is simply not a disease at all; it doesn't exist. The CFS definitions were written in such a manner that CFS becomes like a desert mirage: The closer you approach, the faster it disappears (2006, [Online]).

The only way forward for M.E. patients and all of the diverse patient groups commonly misdiagnosed with 'CFS' (both of which are denied appropriate support, diagnosis and treatment, and may also be subject to serious medical abuse) is that the bogus disease category of 'CFS' must be abandoned.

Every patient deserves the best possible opportunity for appropriate treatment for their illness and for recovery and this process must begin with a correct diagnosis if at all possible. *A correct diagnosis is half the battle won* (Hyde 2006a, 2006b, [Online]) (Hooper 2006, [Online]) (Hyde 2003, [Online]) (Hooper 2003a, [Online]) (Dowsett 2001a, [Online]) (Dowsett 2000, [Online]) (Dowsett 1999a, 1999b, [Online]) (Dowsett n.d., [Online]).

- For more information on why the bogus disease category of 'CFS' must be abandoned see: Who benefits from 'CFS' and 'ME/CFS'?, The misdiagnosis of 'CFS', Why the disease category of 'CFS' must be abandoned and Smoke and Mirrors.
- Those patients misdiagnosed with 'CFS' (and who do not have M.E.) are advised to read the following papers: The Misdiagnosis of 'CFS' and Where to after a 'CFS' (mis)diagnosis?
- *An additional note on 'fatigue':* Just as some M.E. sufferers will experience other non-essential symptoms such as vomiting or night sweats some of the time, but others will not, the same is true of fatigue. The diagnosis of M.E. is determined upon the presence of certain neurological, cognitive, cardiac, cardiovascular, immunological, endocrinological, respiratory, hormonal, muscular, gastrointestinal and other symptoms – the presence or absence of mere 'fatigue' is irrelevant.

What do the terms CFIDS, ME/CFS, CFS/ME, Myalgic Encephalopathy and ME-CFS mean?

When the terms CFS, CFIDS, ME/CFS, CFS/ME, or Myalgic Encephalopathy are used, what is being referred to may be patients with any combination of:

1. Miscellaneous psychological and non-psychological fatigue states (including somatisation disorder).

2. A self limiting post-viral fatigue state or syndrome (e.g. following glandular fever).

3. A mixed bag of unrelated, misdiagnosed illnesses (each of which features fatigue as well as a number of other common symptoms; poor sleep, headaches, muscle pain etc.) including Lyme disease, Multiple Sclerosis, Fibromyalgia, athletes over-training syndrome, depression, burnout, systemic fungal infections (Candida) and even various cancers.

4. Myalgic Encephalomyelitis patients.

The terminology is often used interchangeably, incorrectly and confusingly. However, the DEFINITIONS of M.E. and 'CFS' are very different and distinct, and it is the definitions of each of these terms which is of primary importance. *The distinction must be made between terminology and definitions.*

1. *Chronic Fatigue Syndrome* is an artificial construct created in the US in 1988 for the benefit of various political and financial vested interest groups. It is a mere diagnosis of exclusion (or wastebasket diagnosis) based on the presence of gradual or acute onset fatigue lasting at least 6 months. If tests show serious abnormalities, a person no longer qualifies for the diagnosis, as 'CFS' is 'medically unexplained.' A diagnosis of 'CFS' does not mean that a person has any distinct disease (including M.E.). The patient population diagnosed with 'CFS' is made up of people with a vast array of unrelated illnesses, or with no detectable illness. According to the latest CDC estimates, 2.54% of the population qualifies for a 'CFS' diagnosis. Every diagnosis of 'CFS' can only ever be a misdiagnosis.

2. *Myalgic Encephalomyelitis* is a systemic neurological disease initiated by a viral infection. M.E. is characterised by scientifically measurable damage to the brain, and particularly to the brain stem which results in dysfunctions and damage to almost all vital bodily systems and a loss of normal internal homeostasis.

Substantial evidence indicates that M.E. is caused by an enterovirus. The onset of M.E. is always acute and M.E. can be diagnosed within just a few weeks. M.E. is an easily recognisable distinct organic neurological disease which can be verified by objective testing. If all tests are normal, then a diagnosis of M.E. cannot be correct.

M.E. can occur in both epidemic and sporadic forms and can be extremely disabling, sometimes fatal. M.E. is a chronic/lifelong disease that has existed for centuries. It shares similarities with M.S., Lupus and Polio. There are more than 60 different neurological, cognitive, cardiac, metabolic, immunological and other M.E. symptoms. Fatigue is not a defining or even essential symptom of M.E. People with M.E. would give anything to be only 'fatigued' instead of having M.E. Far fewer than 0.5% of the population has the distinct neurological disease known since 1956 as Myalgic Encephalomyelitis.

The only thing that makes any sense is for patients with M.E. to be studied ONLY under the name Myalgic Encephalomyelitis, and for this term ONLY to be used to refer to a 100% M.E. patient group. The only correct name for this illness – M.E. as per Ramsay/Richardson/Dowsett and Hyde – is Myalgic Encephalomyelitis.

M.E. is not synonymous with 'CFS', nor is it a subgroup of 'CFS'. It is also important that the only terms which are used are those which do have an official and correct World Health Organization classification.

There is no such disease as 'CFS' – the name 'CFS' and the bogus disease category of 'CFS' must be abandoned, along with the use of other vague and misleading umbrella terms such as 'ME/CFS' 'CFS/ME' 'CFIDS, 'Myalgic Encephalopathy' and others, for the benefit of all the patient groups involved.

- For more information on why the bogus disease category of 'CFS' must be abandoned, (along with the use of other vague and misleading umbrella terms such as 'ME/CFS' 'CFS/ME' 'CFIDS' and 'Myalgic Encephalopathy' and others), see: Who benefits from 'CFS' and 'ME/CFS'?, Problems with the so-called "Fair name" campaign: Why it is in the best interests of all patient groups involved to reject and strongly oppose this misleading and counter-productive proposal to rename 'CFS' as 'ME/CFS' and Problems with the use of 'ME/CFS' by M.E. advocates, plus The misdiagnosis of CFS, Why the disease category of 'CFS' must be abandoned and Smoke and Mirrors
- *A note on the current name change proposal:* It is madness to suggest that CFS should be renamed as ME-CFS or CFS/ME or ME/CFS, as some US CFS groups are currently advocating. M.E. and CFS are not the same, only a small percentage of those (mis)diagnosed with CFS qualify for a diagnosis of authentic M.E., the vast majority do not. People with depression, Lyme disease, Candida, etc. do not need to be given an additional misdiagnosis of ME/CFS, they must instead be given a correct diagnosis finally. The fact that some of these patients, and others, may fit the Canadian criteria for 'ME/CFS' does not mean that these patients can be correctly diagnosed with M.E. – as per Ramsay/Richardson/Dowsett and Hyde – nor that these illnesses are the same or 'virtually the same' as M.E. They are not. The Canadian 'ME/CFS' Guidelines and the newer version titled the International Consensus Criteria (ICC) are not accurate M.E. definitions. They are not definitions of M.E. at all. They are both redefinitions of 'CFS' which unscientifically throw in a few facts about M.E. and by doing so unhelpfully worsen the confusion between these two very different entities. For more information see: Canadian Guidelines Review and Testing for M.E.

But isn't the name 'CFS' a big part of the problem?

The reason so many patients are ridiculed, sneered at, belittled, disbelieved, accused of exaggerating or malingering or laziness by medical staff and by friends and family members is not because of the name 'Chronic Fatigue Syndrome'! If 'CFS' had instead been given a neutral name, say 'Reeves' syndrome' or 'Holmes' syndrome,' the problems would still be exactly the same. Vested interest groups – helped in this task immeasurably by the creation of the bogus disease category of 'CFS' – would still be flooding the medical, political and media communities with lies and propaganda which could only have the end result of making patients seem utterly pathetic and undeserving of any respect or sympathy.

What else could anyone think of patients who supposedly have an illness that is mild and short lived, but which some patients pretend is severely disabling because they 'enjoy the sick role'? What else could anyone think about an illness that cannot in any way be proved despite vast sums being spent on tests and that must be taken completely on faith. What else could anyone think about an illness that has seemingly been proven to be psychological or behavioural but where it seems patients would prefer to actually stay ill rather than to admit that they are mentally ill?

Every media article and government press release about 'CFS' is filled with fictional statements which make it very clear in many different ways that the illness has no scientific validity and that the patients do not deserve the same respect as other patient groups, but should be treated with contempt. Patients are not merely wrongly categorized as psychologically ill; it is so much more than that. It is persecution; patients are labelled as malingerers and deviants, and spoken about as if they were beneath contempt and not worthy of even basic respect or medical care, or even any level of kindness or compassion – even from their own friends and family. Whatever 'CFS' had been named, these problems would be the same. There is no such disease as 'CFS' and 'CFS' is merely an artificial entity created for the benefit of financial vested interest groups – that is the real problem, not the name 'CFS.'

What does the term ICD-CFS mean?

The various definitions of 'CFS' *do not* define M.E. Myalgic Encephalomyelitis as an organic neurological disorder as defined at G.93.3 in the World Health Organization's International Classification of Diseases (ICD). The definitions of 'CFS' do not reflect this. The 'CFS' or 'ME/CFS' definitions are not 'watered down' M.E. definitions, as some claim. They are not definitions of M.E. at all.

However, ever since an outbreak of M.E. in the US was given the label 'CFS,' the name/definition 'CFS' has prevailed for political reasons. 'CFS' is widely though wrongly applied to M.E. as well as to other diseases. The overwhelming majority of 'CFS' research does not involve M.E. patients and is not relevant *in any way* to M.E. patients. However, a minuscule percentage of research published under the name 'CFS' clearly does involve a significant number of M.E. patients as it details those abnormalities which are unique to M.E. Sometimes the problematic term 'ICD-CFS' is used in those studies and articles which, while they use the term 'CFS,' do relate to some extent to authentic M.E.

Problems with 'CFS' or so-called 'ICD-CFS' research

The overwhelming majority of 'CFS' research does not involve M.E. patients and is not relevant *in any way* to M.E. patients. A small number of 'CFS' studies refer in part to people with M.E. but it may not always be clear which parts refer to M.E. Unless studies are based on an exclusively M.E. patient group, results cannot be interpreted and are meaningless for M.E. While it is important to be aware of the small amount of research findings that do hold some value for M.E. patients, using the term 'ICD-CFS' to refer to this research is misleading and in many ways just damaging as using terms and concepts like 'ME/CFS' or 'CFS/ME.'

- For further details of the WHO ICD classifications of M.E. and 'CFS' worldwide and why terms such as 'ICD-CFS,' 'ME/CFS' and Myalgic 'Encephalopathy' must be avoided, please see the new paper by patient advocate Lesley Ben entitled: The World Health Organization's International Classification of Diseases (WHO ICD), ME, 'CFS,' 'ME/CFS' and 'ICD-CFS'
- Virtually all of the research which does relate to M.E., at least in part, but which uses the term/concept of 'CFS,' or ME/CFS, or CFIDS etc.— is also contaminated in some way by 'CFS' misinformation. Most often these papers contain a bizarre mix of facts relating to both M.E. and 'CFS.' For more information on some of the most common inaccuracies and 'CFS' propaganda included in this research, see the paper: Putting Research and Articles on M.E. into context and A warning on 'CFS' and 'ME/CFS' research and advocacy

What does define M.E.? What is its symptomatology?

M.E. is a systemic acutely acquired illness, initiated by a virus infection, which is characterised by post encephalitic damage to the brain stem (CNS) — a nerve centre through which many spinal nerve tracts connect with higher centres in the brain in order to control all vital bodily functions. This is always damaged in M.E., hence the name Myalgic Encephalomyelitis.

The CNS is diffusely injured at several levels; these include the cortex, the limbic system, the basal ganglia, the hypothalamus as well as areas of the spinal cord and its appendages. This persisting multilevel CNS dysfunction is undoubtedly both the chief cause of disability in M.E. and the most critical in the definition of the entire disease process.

M.E. represents an acute change in the balance of neuropeptide messengers, and consequently, a resulting loss of the ability of the CNS to adequately receive, interpret, store and recover information which enables it to control vital body functions (cognitive, hormonal, cardiovascular, autonomic and sensory nerve communication, digestive, visual auditory balance etc). It is a loss of normal internal homeostasis. The individual can no longer function systemically within normal limits.

M.E. is primarily neurological, but because the brain controls all vital bodily functions, virtually every bodily system can be affected by M.E. Again, although M.E. is primarily neurological it is also known that the vascular and cardiac dysfunctions seen in M.E. are the cause of many of the symptoms and much of the disability associated with M.E., and that the well-documented mitochondrial abnormalities present in M.E. significantly contribute to both of these pathologies. There is also multi-system involvement of cardiac and skeletal muscle, liver, lymphoid and endocrine organs in M.E. Some individuals also have damage to skeletal and heart muscle.

M.E. symptoms are manifested by virtually all bodily systems including: cognitive, cardiac, cardiovascular, immunological, endocrinological, respiratory, hormonal, gastrointestinal and musculo-skeletal dysfunctions and damage.

M.E. is an infectious neurological disease and represents a major attack on the CNS – and an associated injury of the immune system – by the chronic effects of a viral infection. There is also transient and/or permanent damage to many other organs and bodily systems in M.E.

M.E. affects the body systemically. Even minor levels of physical and cognitive activity, sensory input and orthostatic stress beyond an M.E. patient's individual post-illness limits causes a worsening of the illness, and of symptoms, which can persist for days, weeks, months or even longer. In addition to the risk of relapse, repeated or severe overexertion can also cause permanent damage (e.g. to the heart), disease progression and/or death in M.E.

M.E. is not stable from one hour, day, week or month to the next. It is the combination of the chronicity, the dysfunctions, the instability and the lack of dependability of these functions that creates the high level of disability in M.E. It is also worth noting that of the CNS dysfunctions, cognitive dysfunction is a major disabling characteristic of M.E.

All of this is not simply theory, but is based upon an enormous body of mutually supportive clinical information. These are well-documented, scientifically sound explanations for why patients are bedridden, profoundly intellectually impaired, unable to maintain an upright posture and so on (Chabursky et al. 1992 p. 20) (Hyde 2007, [Online]) (Hyde 2006, [Online]) (Hyde 2003, [Online]) (Hyde 2009) (Dowsett 2001a, [Online]) (Dowsett 2000, [Online]) (Dowsett 1999a, [Online]) (Hyde 1992 pp. x-xxi) (Hyde & Jain 1992 pp. 38 - 43) (Hyde et al. 1992, pp. 25-37) (Dowsett et al. 1990, pp. 285-291) (Ramsay 1986, [Online]) (Dowsett & Ramsay n.d., pp. 81-84) (Richardson n.d., pp. 85-92).

- *What is homeostasis?* Homeostasis is the property of a living organism, to regulate its internal environment to maintain a stable, constant condition, by means of multiple dynamic equilibrium adjustments, controlled by interrelated regulation mechanisms. Homeostasis is one of the fundamental characteristics of living things. It is the maintenance of the internal environment within tolerable limits.

What are some of the symptoms of M.E.?

More than 64 distinct symptoms have been authentically documented in M.E. At first glance it may seem that every symptom possible is mentioned, but although people with M.E. have a lot of different minor symptoms because of the way the central nervous system (which controls virtually every bodily system) is affected, the major symptoms of M.E. really are quite distinct and almost identical from one patient to the next (Hooper & Montague 2001a, [Online]) (Hyde 2006, [Online]). Individual symptoms of M.E. include:

Sore throat, chills, sweats, low body temperature, low grade fever, lymphadenopathy, muscle weakness (or paralysis), muscle pain, muscle twitches or spasms, gelling of the joints, hypoglycaemia, hair loss, nausea, vomiting, vertigo, chest pain, cardiac arrhythmia, resting tachycardia, orthostatic tachycardia, orthostatic fainting or faintness, circulatory problems, opthalmoplegia, eye pain, photophobia, blurred vision, wavy visual field, and other visual and neurological disturbances, hyperacusis, tinnitus, alcohol intolerance, gastrointestinal and digestive disturbances, allergies and sensitivities to many previously well-tolerated foods, drug sensitivities, stroke-like episodes, nystagmus, difficulty swallowing, weight changes, paresthesias, polyneuropathy, proprioception difficulties, myoclonus, temporal lobe and other types of seizures, an inability to maintain consciousness for more than short periods at a time, confusion, disorientation, spatial disorientation, disequilibrium, breathing difficulties, emotional lability, sleep disorders; sleep paralysis, fragmented sleep, difficulty initiating sleep, lack of deep-stage sleep and/or a disrupted circadian rhythm.

Neurocognitive dysfunction may include cognitive, motor and perceptual disturbances. Cognitive dysfunction may be pronounced and may include: difficulty or an inability to speak (or understand speech), difficulty or an inability to read or write or to do basic mathematics, difficulty with simultaneous processing, poor concentration, difficulty with sequencing, and problems with memory including difficulty making new memories, difficulty recalling formed memories and difficulties with visual and verbal recall (e.g. facial agnosia). There is often a marked loss in verbal and performance intelligence quotient (IQ) in M.E. (Bassett 2010, [Online]).

- For a more complete symptom list see: The ultra-comprehensive M.E. symptom list. See also: What it feels like to have M.E.: A personal M.E. symptom list and description of M.E.
- See the Research and Articles section for many hundreds of different articles and medical studies into M.E.

What other features define or characterise M.E.?

What characterises M.E. every bit as much as the individual neurological, cognitive, cardiac, cardiovascular, immunological, endocrinological, respiratory, hormonal, muscular, gastrointestinal and other symptoms is the way in which people with M.E. respond to physical and cognitive activity, sensory input and orthostatic stress;. -n other words, the pattern of symptom exacerbations, relapses and disease progression.

The way the bodies of people with M.E. react to these activities/stimuli post-illness is unique in a number of ways. Along with a specific type of damage to the CNS, this characteristic is one of the defining features of the illness and must be present for a correct diagnosis of M.E. to be made. The main characteristics of the pattern of symptom exacerbations, relapses and disease progression in M.E. include the following:

A. People with M.E. are unable to maintain their pre-illness activity levels. This is an acute, sudden change. M.E. patients can only achieve 50% or less of their pre-illness activity levels.

B. People with M.E. are limited in how physically active they can be but are also limited in similar ways with cognitive exertion, sensory input and orthostatic stress.

C. When a person with M.E. is active beyond their individual physical, cognitive, sensory or orthostatic limits, there is a worsening of various neurological, cognitive, cardiac, cardiovascular, immunological, endocrinological, respiratory, hormonal, muscular, gastrointestinal and other symptoms.

D. The level of physical activity, cognitive exertion, sensory input or orthostatic stress that is needed to cause a significant or severe worsening of symptoms varies from patient to patient, but is often trivial compared to a patient's pre-illness tolerances and abilities.

E. The severity of M.E. waxes and wanes throughout the hour, day, week and month. Adrenaline surges sometimes enable patients to complete tasks they would not usually be able to do, but this comes at the cost of significant relapses or disease progression over time.

F. The worsening of the illness caused by overexertion often does not peak until 24 - 72 hours or more later.

G. The effects of overexertion can accumulate over longer periods of time and lead to disease progression or death.

H. The activity limits of M.E. are not short term: an increase in activity levels beyond a patient's individual limits, even if gradual, causes relapse, disease progression or death.

I. The symptoms of M.E. do not resolve with rest. The symptoms and disability of M.E. are not caused only by overexertion: there is also a base level of illness which can be quite severe even at rest.

J. Repeated overexertion can harm the patient's chances for future improvement in M.E. Patients who are able to avoid overexertion have repeatedly been shown to have the most positive long-term prognosis.

K. Not every M.E. sufferer has 'safe' activity limits within which they will not exacerbate their illness: this is not the case for very severely affected patients.

- For the full-length version of this text and for a full list of references for this text see: <u>The ultra-comprehensive M.E. symptom list</u>.

What causes M.E.?

M.E. expert <u>Dr Byron Hyde</u> explains that:

> [The] prodromal phase is associated with a short onset or triggering illness. This onset illness usually takes the form of either, or any combination, of the following, (a) an upper respiratory illness, (b) a gastrointestinal upset, (c) vertigo and (d) a moderate to severe meningitic type headache. The usual incubation period of the triggering illness is 4-7 days. The second and third phases of the illness are usually always different in nature from the onset illness and usually become apparent within 1-4 weeks after the onset of the infectious triggering illness (1998 [Online]).

Despite popular opinion, (and the vast amount of 'CFS' government and media propaganda which purports to be relevant to M.E. but is not), there is **no** link between contracting M.E. and being a 'perfectionist' or having a 'type A' or over-achieving personality. M.E. **cannot** be caused by a period of long-term or intense stress, trauma or abuse in childhood, becoming run-down, working too hard or not eating healthily. M.E. is not a form of 'burnout' or nervous exhaustion, or the natural result of a body no longer able to cope with long-term stress.

Research also shows that it is simply not possible that M.E. could be caused by the Epstein-Barr virus, any of the herpes viruses (including HHV6), glandular fever/mononucleosis, Cytomegalovirus (CMV), Ross River virus, Q fever, hepatitis, chicken pox, influenza or any of the bacteria which can result in Lyme disease (or other tick-borne bacterial infections). M.E. is also not a form of chemical poisoning.

M.E. is undoubtedly caused by a virus, a virus with an incubation period of 4-7 days. There is also ample evidence that M.E. is caused by the same type of virus that causes Polio: an enterovirus (Hyde 2006, [Online]) (Hyde 2007, [Online]) (Hooper 2006, [Online]) (Hooper & Marshall 2005a, [Online]) (Hyde 2003a, [Online]) (Dowsett 2001a, [Online]) (Hooper et al. 2001, [Online]) (Dowsett 2000, [Online]) (Dowsett 1999a, 1999b, [Online]) (Ryll 1994, [Online]).

- See <u>The outbreaks (and infectious nature) of M.E.</u> section for more information.
- For information on the outrageous hype surrounding the recent XMRV 'CFS' research, please see the <u>XMRV, 'CFS,' and M.E.</u> paper by Sarah Shenk as well as the HFME press release: <u>International M.E. expert disputes that 'CFS' XMRV retrovirus claim has relevance to M.E. patients</u>

Are there outbreaks of M.E.?

One of the most fundamental facts about M.E. throughout its history is that it occurs in epidemics. There is a history of over sixty recorded outbreaks of the illness going back to 1934 when an epidemic of what seemed at first to be Poliomyelitis was reported in Los Angeles. As with many of the other M.E. outbreaks, the Los Angeles outbreak occurred during a local Polio epidemic.

The presenting illness resembled Polio, so for some years the illness was considered to be a variant of Polio and classified as 'Atypical Poliomyelitis' or 'Non-paralytic Polio' (TCJRME 2007, [Online]) (Hyde 1998, [Online]) (Hyde 2006, [Online]). Many early outbreaks of M.E. were also individually named for their locations so we also have outbreaks known as Tapanui flu in New Zealand, Akureyri or Icelandic disease in Iceland, Royal Free Disease in the UK, and so on (TCJRME 2007, [Online]) (Hyde 1998, [Online]).

A review of early M.E. outbreaks found that clinical symptoms were consistent in over sixty recorded epidemics spread all over the world (Hyde 1998, [Online]). Despite the different names being used, these were repeated outbreaks of the same illness. It was also confirmed that the epidemic cases of M.E. and the sporadic cases of M.E. each represented the same illness (Hyde 2006, [Online]) (Dowsett 1999a, [Online]).

M.E. is an infectious neurological disease and represents a major attack on the CNS by the chronic effects of a viral infection. The world's leading M.E. experts, namely Ramsay, Richardson, Dowsett and Hyde, (and others) have all indicated that M.E. is caused by an enterovirus.

The evidence which exists to support the concept of M.E. as an enteroviral disease is compelling (Hyde 2007, [Online]) (Hyde 2006, [Online]). An enterovirus explains the age variation, sex variation, obvious resistance of some family members to the infection and the effect of physical activity — particularly in the early stages of the illness — in creating more long-term/severe M.E. illness in the host (Hyde & Jain 1992a, p. 40).

There is also the evidence that:

- M.E. epidemics very often followed Polio epidemics.
- M.E. resembles Polio at onset.
- Serological studies have shown that communities affected by an outbreak of M.E. were effectively blocked (or immune) from the effects of a subsequent Polio outbreak.
- Evidence of enteroviral infection has been found in the brain tissue of M.E. patients at autopsy (Hyde 2007, [Online]) (Hyde 2006, [Online]) (Hyde 2003, [Online]) (Dowsett 2001a, [Online]) (Dowsett 2000, [Online]) (Dowsett 1999a, 1999b, [Online]) (Hyde 1992 p. xi) (Hyde & Jain 1992 pp. 38 - 43) (Hyde et al. 1992, pp. 25-37) (Dowsett et al. 1990, pp. 285-291) (Ramsay 1986, [Online]).

The US Centres for Disease Control (CDC) placed 'CFS' on its "Priority One, New and Emerging" list of infectious diseases some years ago; a list that also includes Lyme disease, hepatitis C, and malaria' (Gellman & Verillo 1997, p. 19). Despite this, no real research into transmissibility (or more importantly on reducing infection rates) has been done by any government on patients with M.E. (or even 'CFS') despite ample evidence that this is an infectious disease.

There have been many well-documented clusters or outbreaks of the illness, reports of as many as 4.5% of M.E. sufferers contracting the illness immediately after blood transfusions (or after needle-stick injuries involving the blood of M.E. patients) and evidence of the disease spreading through casual contact amongst family members (Carruthers et al. 2003, p.79).

As Dr Elizabeth Dowsett explains: 'The problem we face is that, in spite of overwhelming epidemiological and technical evidence of an infectious cause, the truth is being suppressed by the government and the 'official' M.E. charities as 'too scary' for the general public' (n.d.a, [Online]).

This pretence of ignorance on behalf of government worldwide has had enormous consequences: for example, only in the UK are people with M.E. specifically banned from donating blood. Consequently, the number of people infected with M.E. continues to rise unabated and largely unnoticed by the public.

- See: The outbreaks (and infectious nature) of M.E. page for more information.

Is M.E. difficult to diagnose? What tests can be used to diagnose M.E.?

M.E. is a distinct, recognisable disease entity that is not difficult to diagnose and can in fact be diagnosed relatively early in the course of the disease (within just a few weeks), providing that the physician has some experience with the illness. There is just no other illness that has all the major features of M.E.

Although there is as yet no single test which can be used to diagnose M.E., there are (as with Lupus, Multiple Sclerosis, ovarian cancer and many other illnesses) a *series* of tests which can confirm a suspected M.E. diagnosis. Virtually every M.E. patient will also have various abnormalities visible on physical exam. If all tests are normal, if specific abnormalities are not seen on certain of these tests (e.g. brain scans), then a diagnosis of M.E. cannot be correct (Hyde 2007, [Online]) (Hyde 2006, [Online]) (Hooper et al. 2001, [Online]) (Chabursky et al. 1992, p.22).

As M.E. expert Dr Byron Hyde explains:

> The one essential characteristic of M.E. is acquired CNS dysfunction. A patient with M.E. is a patient whose primary disease is CNS change, and this is measurable. We have excellent tools for measuring these physiological and neuropsychological changes: SPECT, xenon SPECT, PET, and neuropsychological testing (2003, [Online]).

Tests which together can be used to confirm an M.E. diagnosis include:

- SPECT and xenon SPECT scans of the brain
- MRI and PET scans of the brain
- Neurological examination

- Neuropsychological testing (including QEEG scans)
- The Romberg or tandem Romberg test
- Various tests of the immune system (including tests of natural killer cells number and function)
- Insulin levels and glucose tolerance tests
- Sedimentation rate testing (M.E. is one of less than half a dozen diseases which can cause sedimentation rates as low as zero)
- Circulating blood volume tests (which may show a reduced circulating blood volume of up to 50%)
- 24 hour Holter monitor testing (a type of heart monitor)
- Tilt table examination and blood pressure tests
- Exercise testing and chemical stress tests
- Physical exam

These tests are the most critical in the diagnosis of M.E., although various other types of tests are also useful.

Dr Byron Hyde's highly regarded (and TESTABLE) M.E. definition The Nightingale Definition of M.E. makes diagnosis easier than ever before, even for those with no experience with the illness (Hyde 2007, [Online]) (Hyde 2006, [Online]) (Hooper & Marshall 2005a, [Online]) (Hyde 2003, [Online]) (Dowsett 2001a, [Online]) (Dowsett 2000, [Online]) (Hyde 1992 p. xi) (Hyde & Jain 1992 pp. 38 - 43) (Hyde et al. 1992, pp. 25-37) (Dowsett et al. 1990, pp. 285-291) (Ramsay 1986, [Online]) (Dowsett n.d., [Online]) (Dowsett & Ramsay n.d., pp. 81-84) (Richardson n.d., pp. 85-92).

- Objective scientific tests *are* available which can aid in the diagnosis of M.E. and easily prove the severe abnormalities across many different bodily systems seen in M.E. Unfortunately many patients are not given access to these tests. Problems also exist with doctors not being familiar with the abnormalities on testing seen in M.E. and so *misinterpreting* the results of some tests. The problem is not that these tests don't exist, but that doctors – and many patients – are unaware of this information on testing, that it is not generally accepted due to the nefarious influence of political and financial vested interest groups, and that there are overwhelming financial and political incentives for researchers to IGNORE this evidence in favour of the bogus 'CFS' (or 'subgroups of 'ME/CFS') construct. For more information see: Testing for M.E. and Are we just 'marking time?'

How common is M.E.? Who gets M.E. and how?

Although the illness we now know as M.E. has existed for centuries, for much of that time it was a relatively uncommon disease. Following the mass Polio vaccination programs of the 1960s, cases of Polio were greatly reduced and outbreaks of M.E. seemed to be similarly affected. It wasn't until the late 1970s that M.E. began its dramatic increase in incidence worldwide. Over 20 years later, M.E. is a worldwide epidemic of devastating proportions. Many people have died from M.E. and there are now many hundreds of thousands of people severely disabled by this epidemic (TCJRME 2007, [Online]) (Hyde 1992, p. xi).

The main period of infectivity of M.E. peaks at the time just before symptoms appear through to the initial acute phase of the illness (which lasts for several months or in some cases years). M.E. appears to be highly infective but also highly selective. The major mode of infectivity is by an airborne or respiratory route. Modes of transmission are thought to include: casual contact (respiratory), salivary transmission (e.g. kissing), sexual transmission and transmission through blood products (Hyde et al. 1992, pp. 25 - 37). (A recent study of 752 patients found that 4.5% of them – almost one in twenty – had had a blood transfusion days or a week before experiencing acute onset of M.E.) (Carruthers et al. 2003, [Online]) (Hyde et al. 1992, pp. 25 - 37).

M.E. has a similar strike rate (or possibly somewhat higher), to Multiple Sclerosis and is estimated to affect roughly 0.2% of the population. Children and teenagers are also susceptible to the illness and children as young as five have been diagnosed with M.E. (M.E. can occur in children younger than five, but this is thought to be rare.) All ages are affected but most commonly sufferers are under 45 at onset. Women are affected around three times as often as men, a ratio common in autoimmune disorders, although in children the sexes seem to be afflicted equally.

M.E. affects all ethnic and socio-economic groups and has been diagnosed all over the world. There are more than a million M.E. sufferers worldwide (Hooper et al. 2001 [Online]) (Hyde 1992, pp. x - xxi).

- The CDC has recently released vastly inflated estimates for figures affected by 'CFS' but it should be noted that the number of people suffering with sustained fatigue has no more relevance to patients with M.E. than to those with M.S. or AIDS or any other distinct illness. See: <u>More medical 'firsts' from the CDC?</u>

Are there any treatments for M.E.?

There are no easy or quick cures for M.E., nor are any on the horizon – despite a lot of hype about various fairly unpromising 'CFS' research endeavours. Intelligent nutritional, pharmaceutical and other interventions can make a significant difference to a patient's life, however.

Appropriate biomedical diagnostic testing should be done as a matter of course (and repeated regularly) to ensure that the aspects of the illness which are able to be treated *can* be diagnosed, monitored and then treated as appropriate. Testing is also important so that dangerous deficiencies and dysfunctions, which may place the patient at significant risk, are not overlooked (Hooper at al. 2001 [Online]). For specific information on M.E. treatment, the following HFME papers are recommended reading:

- <u>Treating M.E. - The basics</u> and <u>Treating and living with M.E.: Overview</u>
- <u>Finding a good doctor when you have M.E.</u>
- <u>Symptom-based management vs. deep healing in M.E.</u>
- <u>A quick start guide to treating and improving M.E. with aggressive rest therapy, diet, toxic chemical avoidance, medications, supplements and vitamins</u>
- <u>Why research and try treatments when some groups claim an M.E. cure is coming soon?</u>
- <u>What if vitamin/mineral/protocol 'x' didn't work for me?</u>
- <u>Deep healing in M.E.: An order of attack!</u>
- <u>Treating M.E. in the early stages</u>

What is known about M.E. so far?

There is an abundance of research which shows that M.E. is an organic illness which can have profound effects on many bodily systems. These are well-documented, scientifically sound explanations for why patients are bedridden, profoundly intellectually impaired, unable to maintain an upright posture and so on. More than **a thousand** good articles now support the basic premises of M.E. Autopsies have also confirmed such reports of bodily damage and infection (Hooper & Williams 2005a, [Online]).

Many different organic abnormalities have been found in M.E. patients (in peer reviewed research).

It is known that M.E. is:

1. An acute onset (biphasic) epidemic or endemic infectious disease process
2. An autoimmune disease (with similarities to Lupus)
3. An infectious neurological disease, affecting adults and children
4. A disease which involves significant (and at times profound) cognitive impairment/dysfunction
5. A persistent viral infection (due to an enterovirus; the same type of virus which causes Poliomyelitis and post-Polio syndrome)
6. A diffuse and measurable injury to the vascular system of the CNS.
7. A CNS disease with similarities to M.S.
8. A variable (but always serious) diffuse, acquired brain injury
9. A systemic illness (associated with organ pathology; particularly cardiac)
10. A vascular disease
11. A cardiovascular disease
12. A type of cardiac insufficiency
13. A mitochondrial disease
14. A metabolic disorder
15. A musculo-skeletal disorder
16. A neuroendocrine disease
17. A seizure disorder
18. A sleep disorder

19. A gastrointestinal disorder
20. A respiratory disorder
21. An allergic disorder
22. A pain disorder
23. A life-altering disease
24. A chronic or lifelong disease associated with a high level of disability
25. An unstable disease: from one hour/day/week or month to the next
26. A potentially progressive or fatal disease

M.E. affects every cell in the body and almost every bodily system (Hyde 2007, [Online]) (Hooper et al. 2001, [Online]) (Cheney 2007, [video recording]) (Ramsay 1986, [Online]).

- For more information see the General articles and research overviews section. See also articles by: Dr Elizabeth Dowsett and Dr Byron Hyde.

Is there a legitimate scientific debate about whether or not M.E. is a 'real' neurological disease?

Despite popular opinion there simply is no legitimate scientifically motivated debate about whether or not M.E. is a 'real' neurological illness or not, or whether it has a biological basis.

The psychological or behavioural theories of M.E. and claims that M.E. is just another term for 'CFS' are no more scientifically viable than theories of a flat earth. They are pure fiction.

Are there any somewhat similar medical conditions?

There are a number of post-viral fatigue states or syndromes which may follow common infections such as mononucleosis/glandular fever, hepatitis, Q fever, Ross River virus and so on. M.E. is an entirely different condition to these self-limiting fatigue syndromes however, and it is *not* caused by the Epstein Barr virus or any of the herpes or hepatitis viruses. People suffering with any of these post-viral fatigue syndromes do not have M.E.

M.E. does have some limited similarities – to varying degrees – to illnesses such as Multiple Sclerosis, Lupus, post-Polio syndrome, Gulf War Syndrome and chronic Lyme disease, and others. But this does not mean that they represent the same etiological or pathobiological process. They do not.

M.E. is a distinct neurological illness with a distinct onset, symptoms, aetiology, pathology, response to treatment, long and short term prognosis, and World Health Organization classification (G.93.3) (Hyde 2006, [Online]) (Hyde 2007, [Online]) (Hooper 2006, [Online]) (Hooper & Marshall 2005a, [Online]) (Hyde 2003a, [Online]) (Dowsett 2001a, [Online]) (Hooper et al. 2001, [Online]) (Dowsett 2000, [Online]) (Dowsett 1999a, 1999b, [Online])

- See M.E. and other illnesses for more information. See also: M.E. vs. M.S.: Similarities and differences.

How well is research into M.E. research funded by government?

Governments around the world are currently spending $0 a year on M.E. research. Considering the severity of the illness and the vast numbers of patients involved, this is a worldwide disgrace.

- See Putting research and articles on M.E. in context and A warning on 'CFS' and 'ME/CFS' research and advocacy for more information about research into M.E. and the challenges involved. See the Donations page on the HFME website to make a donation towards M.E. research and advocacy.

Abuse and M.E.

Two of the most common interventions people with M.E. are encouraged to participate in are cognitive behavioural therapy (CBT) and graded exercise therapy (GET).

However, despite the misleading claims to the contrary made by various vested interest groups, no evidence exists which demonstrates that CBT and GET are appropriate, effective or safe treatments for M.E. patients. Studies by these groups (and others) involving miscellaneous psychiatric and non-psychiatric 'fatigue' sufferers, and their positive response to these treatments, have no more relevance to M.E. sufferers than they do to patients with Multiple Sclerosis, diabetes or any other illness. Patients with M.E. are routinely being prescribed these treatments on what amounts to a random basis medically.

As (very bad) luck would have it, graded exercise programs are probably the single most inappropriate 'treatment' that an M.E. sufferer could be encouraged to undertake. Permanent damage may result, as well as disease progression. Patient accounts of leaving exercise programs much more severely ill than when they began them are common: some end up wheelchair-bound, bed-bound or requiring hospitalisation in intensive care or cardiac care units. The damage caused is often severe and either long-term or permanent: some patients are still dealing with the effects of inappropriate advice to exercise five, ten or more YEARS afterwards, and for some patients this damage is permanent. Sudden deaths have also been reported in a small percentage of M.E. patients following exercise.

CBT and GET are at best useless and at worst extremely harmful for M.E. patients. Despite this, these 'treatments' are regularly recommended for people with M.E., who are assured that they are completely safe. Patient participation is not always voluntary. Many M.E. patients have been treated as psychiatric patients against their will (or against their parents' will in the case of children with M.E.). In some cases it is a condition of receiving medical insurance or government welfare entitlements that M.E. patients first undergo 'rehabilitation', including CBT and GET programs, particularly in the UK.

If a prescription drug had anything like the appalling track record exercise has with people with M.E. (or even a small fraction of it, even 2%) it would be a worldwide scandal. The drug would be immediately banned, there would be some form of inquiry and serious criminal charges may well be laid. Yet the rate of people with M.E. encouraged or even *forced* to exercise continues to rise, and with the full support of governments. This is despite the fact that legitimate research clearly shows that along with the huge risk involved, it has a zero percent chance of providing any benefit to people with authentic M.E. That this can be allowed to go on in such a supposedly enlightened day and age as ours defies belief.

It is also of great concern that so many M.E. patients are ONLY offered 'treatments' such as CBT and GET, while access to even basic appropriate medical care is withheld. Of the 30% of patients who are severely affected by the illness (and are bed-bound and housebound), the majority have no contact with the health service at all as they are seldom able to obtain house calls (Dunn 2005, [Online]). Many sufferers are also refused the basic welfare support to which they are entitled.

Thus a significant percentage of very physically ill and vulnerable M.E. patients are simply left to suffer and die at home without any medical care, welfare or social support (Hooper 2003a, [Online]).

- These brief comments on the effects of CBT and GET are taken from the more detailed paper: The effects of CBT and GET on patients with M.E., see this paper for more information.
- A recent example of an M.E. sufferer being taken into psychiatric care against their will is the case of Sophia Mirza, in the UK. Tragically, Sophia died of her illness after being wrongly sectioned under the Mental Health Act. Sophia was severely ill. and bedbound but she was refused even basic medical care, and this is believed to have contributed greatly to her death. For more information on this tragic case and entirely avoidable death, see: Inquest Implications, Civilization: Another word for barbarism by Gurli Bagnall and The Story of Sophia and M.E.
- For more information about forced exercise 'treatments' see the 100+ page CBT and GET Database. See also Comments on the 'Lightning Process' scam and other related scams aimed at M.E. patients

Is it only M.E. patients who are negatively affected by the bogus creation of 'CFS'?

If only. Vast numbers of patients from all sorts of varied patient groups misdiagnosed as' CFS' are also denied appropriate diagnosis and treatment, and may routinely be subjected to inappropriate psychological interventions such as CBT and GET. The 'CFS' insurance company scam also impacts negatively on doctors and the general public. The only groups which gain from the 'CFS' confusion are insurance companies and various other organisations and corporations, including the government, which have a vested financial interest in how these patients are treated.

- For more information see: The misdiagnosis of 'CFS' and Who benefits from 'CFS' and 'ME/CFS'?

How severe is M.E.?

Although some people do have more moderate versions of the illness, symptoms are extremely severe for at least 30% of the people who have M.E., significant numbers of whom are housebound and bedbound.

Dr Paul Cheney stated before a US FDA Scientific Advisory Committee:

> I have evaluated over 2,500 cases. At worst, it is a nightmare of increasing disability with both physical and neurocognitive components. The worst cases have both an M.S.-like and an AIDS-like clinical appearance. We have lost five cases in the last six months. 80% of cases are unable to work or attend school. We admit regularly to hospital with an inability to care for self (Hooper et al. 2001 [Online]).

M.E. patients have been found to experience greater functional severity than the studied patients with heart disease, virtually all types of cancer, and all other chronic illnesses. In the 1980s Mark Loveless, an infectious disease specialist and head of the AIDS Clinic at Oregon Health Sciences University which also cared for patients with M.E., found that M.E. patients whom he saw had far lower scores on the Karnofsky performance scale than his HIV patients even in the last week of their life. He testified that an M.E. patient, 'feels effectively the same every day as an AIDS patient feels two weeks before death' (Hooper & Marshall 2005a, [Online]).

But in M.E., this extremely high level of illness and disability is not short-term. It does not always lead to death and it can instead continue uninterrupted for **decades**.

- For more information on severe M.E. see The severity of M.E. and M.E. fatalities and Why patients with severe M.E. are housebound and bedbound.
- Patients with M.E. may also find the following papers useful: Adjusting personal care tasks for the M.E. patient and The HFME M.E. ability and severity scale checklist
- If you would like a friend or family member to be included in the HFME M.E. memorial list, please see the HFME memorial lists page for contact details, and for further information.
- It should also be noted that even those patients with moderate M.E. are far more affected than many patients with a variety of other illnesses. Of course severe M.E. is even worse, but moderate M.E. can also cause significant symptoms and a relatively higher level of disability and suffering than many other illnesses.

Recovery from M.E.

M.E. patients who are given advice to rest in the early stages of the illness, and who avoid overexertion thereafter, have repeatedly been shown to have the most positive long-term prognosis.

As M.E. expert Dr Melvin Ramsay explains:

> The degree of physical incapacity varies greatly, but the [level of severity] is directly related to the length of time the patient persists in physical effort after its onset; put in another way, **those patients who are given a period of enforced rest from the onset have the best prognosis**. Since the limitations which the disease imposes vary considerably from case to case, the responsibility for determining these rests upon the patient. Once these are ascertained the patient is advised to fashion a pattern of living that comes well within them (1986, [Online]).

M.E. can be progressive, degenerative (change of tissue to a lower or less functioning form, as in heart failure), chronic, or relapsing and remitting. Some patients experience spontaneous remissions — albeit most often at a greatly reduced level of functioning compared to pre-illness — and such patients remain susceptible to relapses for the remainder of their lives. M.E. is a chronic/life-long disability where relapse is always possible. Cycles of severe relapse are common, as are further symptoms developing over time. Around 30% of cases are progressive and degenerative and sometimes M.E. is fatal. As Dr Elizabeth Dowsett writes:

> After a variable interval, a multi-system syndrome may develop, involving permanent damage to skeletal or cardiac muscle and to other "end organs" such as the liver, pancreas, endocrine glands and lymphoid tissues,

signifying the further development of a lengthy chronic, mainly neurological condition with evidence of metabolic dysfunction in the brain stem. Yet, stabilisation, albeit at a low level, can still be achieved by appropriate management and support. The death rate of 10% occurs almost entirely from end-organ damage within this group (mainly from cardiac or pancreatic failure) (2001a, [Online]).

Clearly, many people with M.E. are significantly or severely disabled. But what is so tragic about this high level of suffering is that so much of it is needless. The appropriate support (financial, medical and practical) can do much to prevent the physical, occupational and deterioration in quality of life for M.E. patients and can stabilise the illness (Dowsett 2002b, [Online]).

Many deaths from M.E. could have been prevented if only those patients had been given a basic level of support and care made available to patients with illnesses with comparable care needs such as M.S. and Motor Neurone Disease.

- The 3 Part M.E. Ability and Severity Scale can be used to measure M.E. severity over time.
- For information on adrenaline surges in M.E., and the different order in which certain bodily systems may be affected by M.E. (and by overexertion), see the Dr Cheney section in The effects of CBT and GET on patients with M.E. or The importance of avoiding overexertion in M.E. (*Note that Dr Cheney does unfortunately mix M.E. and 'CFS' information and so cannot be considered an M.E. expert, as such.*)

Conclusion

Certain groups and individuals are benefiting enormously from this fraudulent artificial 'CFS' construct.

To say that these groups and individuals always believe what they are saying and that it is based on science or reality is ridiculous. To say that it is merely a misunderstanding or a mistake is equally ridiculous. The 'CFS' construct is a complete fiction, and exists purely because it is so financially and politically beneficial to a number of powerful groups.

The artificial 'CFS' construct is no more a scientifically accurate description of M.E. than it is a scientifically accurate description of M.S., Lupus or Polio. This pretence of ignorance about M.E. and about the reality of 'CFS', particularly by governments, has had devastating consequences for people with M.E. – as well as all of those with non-M.E. illnesses who are misdiagnosed as having 'CFS' – and has also meant that the number of M.E. sufferers continues to rise unabated and largely unrecognised. The general public worldwide, including sufferers themselves, has been lied to repeatedly about the reality of M.E.

The continuing, decades old, systemic abuse and neglect of the million or more people with M.E. worldwide has to stop. M.E. and' CFS' are *not* the same. Concepts such as 'ME/CFS,' 'CFS/ME,' Myalgic 'Encephalopathy' and 'CFIDS' are also unhelpful, unscientific and only add to the obfuscation.'CFS' is merely a scam invented by insurance companies motivated by profit without regard for truth or ethics. These groups are acting without any regard for the extreme suffering and avoidable deaths they are causing. These groups are acting criminally. The scam is tissue thin and very easily discovered if one merely takes the time to look at the evidence.

Why is almost nobody doing this? Why is the world letting these groups get away with such a heinous scam and such appalling abuse on a massive scale? Why isn't the world caring enough or smart enough or gutsy enough to see through these slick, well-funded misinformation campaigns, and to act? How can this be, when the lies are so flimsy and scientifically laughable? Have we learned nothing from the devastating corporate cover-ups of the truth about tobacco and asbestos in our recent past? Where is the World Health Organisation? Where are our human rights groups? Where is our media? Where are our uncompromising investigative journalists?

Will it take another 20 years? How much more extreme do the suffering and abuse have to be? How many more hundreds of thousands of children and adults worldwide have to be affected? How many more patients will have to die needlessly before something is finally done? How much longer will we leave the fox in charge of the hen house?

It's insupportable.

Where do we go from here?

Sub-grouping different types of 'CFS,' refining the bogus 'CFS' definitions further or renaming 'CFS' with some variation on the term M.E. would achieve nothing and create yet more confusion and mistreatment. The problem is not that 'CFS' patients are being mistreated as psychiatric patients; some of those patients misdiagnosed with' CFS' actually *do* have psychological illnesses.

There is no such distinct disease as 'CFS' – that is the entire issue, and the vast majority of patients misdiagnosed with' CFS' *do not* have M.E. and so have no more right to that term than to 'cancer' or 'diabetes.' The only way forward, for the benefit of society and every patient group involved, is that:

1. The bogus disease category of 'CFS' must be abandoned completely. Patients with fatigue (and other symptoms) caused by a variety of different illnesses need to be diagnosed correctly with these illnesses if they are to have any chance of recovery, and not given a meaningless 'CFS' misdiagnosis. Patients with M.E. need this same opportunity. Each of the patient groups involved must be correctly diagnosed and treated as appropriate, based on legitimate and unbiased scientific evidence involving the SAME patient group.

2. The name Myalgic Encephalomyelitis must be fully restored (to the exclusion of all others) and the World Health Organization classification of M.E. (as a distinct neurological disease) must be accepted and adhered to in all official documentations and government policy. As Professor Malcolm Hooper explains:

> The term Myalgic Encephalomyelitis was first coined by Ramsay and Richardson and has been included by the World Health Organisation (WHO in their International Classification of Diseases (ICD), since 1969. The current version ICD-10 lists M.E. under G.93.3 - neurological conditions. It cannot be emphasised too strongly that this recognition emerged from meticulous clinical observation and examination (2006, [Online]).

3. People with M.E. must immediately stop being treated as if they are mentally ill or suffer with a behavioural illness; as if their physical symptoms do not exist or can be improved with 'positive thinking' and exercise, or be mixed in with various 'fatigue' sufferers or patients with any other illness than authentic Myalgic Encephalomyelitis. People with M.E. must also be given access to basic medical care, financial support and other appropriate services (including funding for legitimate M.E. research) on an equal level to that which is available for those with comparable illnesses (e.g. M.S. or Lupus). The facts about M.E. must be taught to medical students, and included in mainstream medical journals.

- See On the Name Myalgic Encephalomyelitis for more information on the evidence for inflammation of the brain and spinal cord in M.E. and other issues surrounding the name Myalgic Encephalomyelitis.

What can you do to help?

Unlike people with HIV/AIDS, people with M.E. do not have an initial period of their illness where they are only mildly affected. M.E. is severely disabling even in the first week of illness. People with M.E. are almost all far too ill to stage protests, rallies or marches. Many with M.E. cannot even read enough to be able to understand what is happening, and are not even aware that high quality scientific information on M.E. exists and that supporting the various 'CFS' and 'ME/CFS' faux 'advocacy' groups is counter-productive in the extreme.

Almost all so-called patient advocacy groups worldwide have sold patients out to the highest bidder and are now *actively collaborating* with our abusers. These groups are no longer advocates for patients with M.E. – indeed they are working directly AGAINST the interest of people with M.E. These groups also do not help all those misdiagnosed with 'CFS', who do not have M.E. The media too has sold-out and betrayed M.E. patients. People with M.E. have only a tiny minority of the medical, scientific, legal and other potentially supporting professions, as well as the public, on their side.

The Committee for Justice and Recognition of Myalgic Encephalomyelitis explains:

> There is no immunity to M.E. The next victim of this horrible disease could be your sister, your friend, your brother, your grandchildren, your neighbour [or] your co-worker. M.E. is an infectious disease that has become

a widespread epidemic that is not going away. We must join together, alert the public and demand action (2007, [Online]).

That is what is needed – people power. Educated people power. For people from all over the world to stand up for M.E. Individual physicians, journalists, politicians, human rights campaigners, patients, families and friends of patients and the public, whether they are affected yet by M.E. or not, must stand up for the truth. That is the only way change will occur— through education and people simply refusing to accept what is happening any more.

Yes, there are powerful and immensely wealthy vested interest groups out there, who will fight the truth every step of the way, but we have science, reality and ethics on our side and those are also very powerful. However, for this to be of any use to us, we must first make ourselves aware of the facts *and then use them.* **So what you can do to help is to PLEASE spread the truth about M.E. and try to expose the lie of 'CFS.'**

You can also help by NOT supporting the bogus concepts of 'CFS,' 'ME/CFS,' 'subgroups of ME/CFS,' 'CFS/ME,' 'CFIDS' and Myalgic 'Encephalopathy.' Do not give public or financial help or support to groups which promote these harmful and unscientific concepts or which equate M.E. with 'CFS.'

The abuse and neglect of so many seriously ill people on such an industrial scale is truly inhumane and has already gone on for far too long. People with M.E. desperately need your help.

References

All of the information concerning Myalgic Encephalomyelitis on this website is fully referenced and has been compiled using the highest quality resources available, produced by the world's leading M.E. experts. More experienced and more knowledgeable M.E. experts than these – Dr Byron Hyde and Dr Elizabeth Dowsett in particular – do not exist.

Between Dr Byron Hyde and Dr Elizabeth Dowsett, and their mentors the late Dr John Richardson and Dr Melvin Ramsay (respectively), these four doctors have been involved with M.E. research and M.E. patients for well over 100 years collectively, from the 1950s to the present day. Between them they have examined more than 15 000 individual (sporadic and epidemic) M.E. patients, as well as each authoring numerous studies and articles on M.E., and books (or chapters in books) about M.E. Again, more experienced, more knowledgeable and more credible M.E. experts than these simply do not exist.

This paper is intended to provide a brief summary of the most important facts of M.E. It has been created for the benefit of those people without the time, inclination or ability to read each of the far more detailed and lengthy references created by the world's leading M.E. experts. The original documents used to create this paper are essential additional reading, however, for any physician (or anyone else) with a real interest in Myalgic Encephalomyelitis. For a full reference list please see the References page of this book.

Acknowledgments
Thanks to Peter Bassett, Emma Searle and Lesley Ben for editing this paper.

Permission is given for each of the individual papers in this book to be freely redistributed by email or in print for any genuine not-for-profit purpose provided that the entire text, including this notice and the author's attribution, is reproduced in full and without alteration.

Disclaimer:
HFME does not dispense medical advice or recommend treatment, and assumes no responsibility for treatments undertaken by visitors to the site or readers of any of the HFME books. It is a resource providing information for education, research and advocacy only. Please consult your own health-care provider regarding any medical issues relating to the diagnosis or treatment of any medical condition.

Relevant quotes
'The problem with fatigue is that it is neither specific, definable nor scientifically measurable. Fatigue is both a normal and a pathological feature of every day life. Every normal person gets fatigued. Fatigue is a

common feature of much major psychiatric disease and major medical disease. Since fatigue is such an integral part of many illnesses, by calling fatigue the primary characteristic, the authors necessitated the elimination of hundreds of other diseases. To truly follow the criteria set out by the CDC definition probably makes 'CFS' the most expensive illness to investigate of any known disease. Fatigue is not an object, it is simply a modifier in search of a noun. Also, taking fatigue as the flagship symptom of a disease not only bestows the disease with a certain Rip Van Winkle humour, but it removes the urgency of the fact that the majority of M.E. symptoms are in effect CNS symptoms. M.E. represents a major attack on the CNS by the chronic effects of a viral infection.'
DR BYRON HYDE IN 'THE CLINICAL AND SCIENTIFIC BASIS OF M.E. P 11-12

'Western newspapers and magazines are packed with trivia, television news is concealing the reality of what is happening… and investigative journalism has virtually died a death. [But] what is the point of democracy if you keep the citizens in a state of semi-ignorance?'
VETERAN ACTIVIST, PROTESTER AND AUTHOR TARIQ ALI

In the mid 1980s, the incidence of M.E. had increased by some seven times in Canada and the UK, while in the USA a major outbreak at Lake Tahoe (wrongly ascribed at first to a herpes virus) led to calls for a new name and new definition for the disease, more descriptive of herpes infection. This definition based on "fatigue" (a symptom common to hundreds of diseases and to normal life, but not a distinguishing feature of myalgic encephalomyelitis) was designed to facilitate research funded by the manufacturers of new anti-herpes drugs. However, a "fatigue" definition (which also omits any reference to children) has proved disastrous for research in the current decade.
RESEARCH INTO M.E. 1988 - 1998 TOO MUCH PHILOSOPHY AND TOO LITTLE BASIC SCIENCE! BY DR ELIZABETH DOWSETT

'Fatigue, which is simply one of the common features of healthy life and disease, neither defines M.E. nor clarifies the illness. The term 'fatigue' does cause disparagement to those who study this serious debilitating illness and those who suffer from it.'
DR BYRON HYDE

'There are actually 30 well documented causes of 'chronic fatigue'. To say that M.E. is a 'subset' of CFS is just as ridiculous as to say it is a 'subset' of diabetes or Japanese B encephalitis or one of the manifestly absurd psychiatric diagnosis, such as, 'personality disorder' or 'somatisation.'''
DR ELIZABETH DOWSETT

'M.E. [is] a loss of the ability of the central nervous system (CNS) to adequately receive, interpret, store and recover information. This dysfunction also results in the inability of the CNS to consistently programme and achieve normal smooth end organ response. [It is a] loss of normal internal homeostasis. The neurochemical homeostatic events continue to be employed uselessly and to the detriment of the organism. This modulatory biochemical complex, biologically derived over the millennium to assist the organism, destabilises the autonomic neuronal outflow and the individual can no longer function systemically within normal limits.'
DR BYRON HYDE

'Do not for one minute believe that CFS is simply another name for Myalgic Encephalomyelitis (M.E.). It is not. The CDC definition is not a disease process.'
DR BYRON HYDE 2006

'Myalgic Encephalomyelitis is a clearly defined disease process. CFS by definition has always been a syndrome. M.E. and CFS are not the same.'
DR BYRON HYDE 2006

'Thirty years ago when a patient presented to a hospital clinic with unexplained fatigue, any medical school physician would search for an occult malignancy, cardiac or other organ disease, or chronic infection. The concept that there is an entity called chronic fatigue syndrome has totally altered that essential medical guideline. Patients are now being diagnosed with CFS as though it were a disease. It is not. It is a patchwork of symptoms that could mean anything.'
DR BYRON HYDE 2003

CHAPTER THREE
More information about severe M.E.

This chapter includes the following paper:

Why patients with severe M.E. are housebound and bedbound

Knowledge of some of the basics of how M.E. affects the body and the limitations of each patient are vital if you provide care for someone with M.E. or even if you make comments or have any type of input into the way the disease is managed, in order to avoid unnecessary suffering and disability.

This paper provides a brief overview of this topic for carers, doctors or hospital staff, as well as friends and family members of patients.

Why severe M.E. patients are housebound and bedbound

 Knowledge of some of the basics of how Myalgic Encephalomyelitis (M.E.) affects the body and the limitations of each patient are vital if you provide care for someone with M.E. This knowledge is also necessary even if you make comments or have any type of input into the way the disease is managed, in order to avoid additional unnecessary suffering and disability. This is so important with M.E. because inappropriate care, comments and advice or pressure for M.E. patients to do certain things (even if well intentioned) can have serious consequences for the patients in the short term and the long term, or even permanently.

This paper provides a brief overview of this topic for friends and family members, and also for carers, doctors or hospital staff.

Why are some severely affected M.E. patients housebound?

This is a question that severe M.E. patients are sometimes asked. The short answer to this question is:

A. They are simply too ill and disabled to leave the house. This task is physically impossible for them due to the severity of their illness, or:

B. They are physically able to leave the house, but it would be unwise for them to do so. In the short term this type of overexertion causes even more severe suffering than is already experienced daily (and may already be at an unbearable level). Even worse, this extreme additional loss of quality of life and ability can and does persist for a long time afterward.

It is very common for severely affected patients to spend 2 months, 6 months, 12 months or even several years or longer recovering from a hospital trip (etc.). For example, some patients still have not regained their previous low-level of health 2 or 4 years after a trip to hospital. Some never do recover, and for some patients the overexertion is so severe as to be fatal.

Severe overexertion also ruins a patient's chances for significant (or any) future recovery, and can cause permanent physical damage.

Severely affected M.E. patients may also sometimes be asked questions such as:

- 'Why are you bedbound, or wheelchair-bound?
- 'Why are you almost completely housebound or bedbound?'
- 'Why have you had to stop studying or working?'
- 'Why can't you do all the tasks of daily living for yourself?'
- 'Why can't you use the phone, or watch TV?'

The answer to each of these questions is the same, it's just a difference of degree. Some tasks are physically impossible for some sufferers, and others are possible but unwise. Sometimes tasks can be done in a controlled way, and limited as to frequency and/or duration. In other words, the activities need to be carefully 'rationed.'

That is really all there is to it. A person with M.E. doesn't do certain things they would like to, because they are either too ill to do them, or because it would reduce their ability and quality of life for months or years afterward. They may lose any chance at significant recovery by pushing themselves to do something that their severely damaged bodies can't cope with.

That is the short answer. If you'd like more detail on all of these points, and some more M.E.-specific medical information and treatment and management guidelines, then please read on.

What is M.E.? How does it affect the body?

Myalgic Encephalomyelitis is a debilitating neurological (CNS) disease which has been recognised by the World Health Organisation since 1969 as a distinct organic neurological disorder. It can occur in both epidemic and sporadic forms and over 60 outbreaks of M.E. have been recorded worldwide since 1934.

M.E. is an acute onset neurological disease initiated by a virus (an enterovirus) with multi system involvement which is characterised by post encephalitic damage to the brain stem (hence the name 'Myalgic Encephalomyelitis'). M.E. is similar in a number of significant ways to diseases such as Multiple Sclerosis (M.S.), Lupus and Polio. M.E. can be extremely disabling; at least 30% of M.E. sufferers are severely affected and are almost completely (or completely) housebound and/or bedbound. Children as young as five can get M.E., as well as adults of all ages. M.E. has a similar strike-rate to M.S. and is a (potentially fatal) chronic/lifelong illness.

M.E. is primarily neurological, but because the brain controls all vital bodily functions virtually every bodily system can be affected by M.E. Although M.E. is primarily neurological it is also known that the vascular and cardiac dysfunctions seen in M.E. are also the cause of many of the symptoms and much of the disability associated with M.E., and that the well-documented mitochondrial abnormalities present in M.E. significantly contribute to both of these pathologies. There is also multi-system involvement of cardiac and skeletal muscle, liver, lymphoid and endocrine organs in M.E.

M.E. symptoms are manifested by virtually all bodily systems including: cognitive, cardiac, cardiovascular, immunological, endocrinological, respiratory, hormonal, gastrointestinal and musculo-skeletal dysfunctions and damage. M.E. affects the brain, the heart, almost every bodily system and every cell of the body. One of the defining features of M.E. is an inability to maintain homeostasis.

All of this is not simply theory, but is based upon an enormous body of mutually supportive clinical information. These are well-documented, scientifically sound explanations for why patients are housebound or bedridden, profoundly intellectually impaired, unable to maintain an upright posture and so on (Chabursky et al. 1992 p. 20) (Hyde 2007, [Online]) (Hyde 2006, [Online]) (Hyde 2003, [Online]) (Dowsett 2001a, [Online]) (Dowsett 2000, [Online]) (Dowsett 1999a, 1999b, [Online]) (Hyde 1992 pp. x-xxi) (Hyde & Jain 1992 pp. 38 - 43) (Hyde et al. 1992, pp. 25-37) (Dowsett et al. 1990, pp. 285-291) (Ramsay 1986, [Online]) (Dowsett & Ramsay n.d., pp. 81-84) (Richardson n.d., pp. 85-92).

What all of this means in practice is that patients with M.E. have to be very careful with, or limit:

- Physical activity
- Cognitive activity
- Sensory input (exposure to light, noise, movement and vibration), and
- Orthostatic stress (maintaining an upright posture)

The main characteristics of the pattern of symptom exacerbations, relapses and disease progression (and so on) in M.E. include:

A. People with M.E. are unable to maintain their pre-illness activity levels. This is an acute (sudden) change. M.E. patients can only achieve 50%, or less, of their pre-illness activity levels post-M.E.

B. People with M.E. are limited in how physically active they can be but they are also limited in similar way with; cognitive exertion, sensory input and orthostatic stress.

C. When a person with M.E. is active beyond their individual (physical, cognitive, sensory or orthostatic) limits this causes a worsening of various neurological, cognitive, cardiac, cardiovascular, immunological, endocrinological, respiratory, hormonal, muscular, gastrointestinal and other symptoms.

D. The level of physical activity, cognitive exertion, sensory input or orthostatic stress needed to cause a significant or severe worsening of symptoms varies from patient to patient, but is often trivial compared to a patient's pre-illness tolerances and abilities.

E. The severity of M.E. waxes and wanes throughout the hour, day, week and month. Adrenaline surges sometimes enable patients to complete tasks they would not usually be able to do, but this comes at the cost of significant relapses or disease progression over time.

F. The worsening of the illness caused by overexertion often does not peak until 24 - 72 hours (or more) later.

G. The effects of overexertion can accumulate over longer periods of time and lead to disease progression, or death.

H. The activity limits of M.E. are not short term: a gradual (or sudden) increase in activity levels beyond a patient's individual limits can only cause relapse, disease progression or death in patients with M.E.

I. The symptoms of M.E. do not resolve with rest. The symptoms and disability of M.E. are not just caused by overexertion; there is also a base level of illness which can be quite severe even at rest.

J. Repeated overexertion can harm the patient's chances for future improvement in M.E. M.E. patients who are able to avoid overexertion have repeatedly been shown to have the most positive long-term prognosis.

K. Not every M.E. sufferer has 'safe' activity limits within which they will not exacerbate their illness; this is not the case for the very severely affected.

In short, if patients with M.E. exceed their individual physical, cognitive, orthostatic and other limits, they will experience some combination of the following:

- A mild-severe (acute or delayed) worsening of <u>one or more symptoms</u> for hours, days or longer afterward
- A mild-severe (acute or delayed) worsening of <u>virtually every symptom</u> for hours, days or longer afterward
- A severe (acute or delayed) worsening of the <u>base level of illness/disability</u> for hours/ weeks/ months or even years afterward, or
- A <u>permanent</u> worsening of the <u>base level of illness/disability</u> (i.e. permanent physical damage is caused and chances for significant recovery are adversely affected or lost entirely. Painstaking gains made slowly over many months or years may also be lost.)

It is also important to be aware that repeated or severe overexertion can result in the death of the M.E. patient. (Death in M.E. is most often caused by heart failure or multiple organ failure.) (Bassett, 2010, [Online])

For these reasons, it is vitally important that patients are allowed to judge *for themselves* how much activity it is safe and wise for them to attempt. Patients are the best judges of their own limits, and patients' judgements must not be over-ruled. Patients should never be advised, encouraged or forced to be more active than their severely damaged bodies can handle; these decisions cannot safely or ethically be made by any third party.

What are the problems for severe M.E. patients being out of their bed or home?

"How is the M.E. patient being overexerted and made more ill if they are transported somewhere while lying down?," or " how can just a few minutes or hours out of bed possibly make the patient more ill long-term?" a healthy person might ask.

It is common for people dealing with M.E. patients to pay close attention to the fact that a patient with M.E. has to limit physical overexertion, but to not fully understand that excessive sensory input and cognitive exertion and other factors can make the patient just as ill as excess physical activity. These factors are also much harder to minimise. For example:

- It is impossible to avoid additional <u>cognitive stimulus</u> during a trip out of the house. Whether it is looking at new environments, or having to listen to speech or being asked to answer questions and make decisions or just being asked to speak at all, all of these things can be unbearable for the severe M.E. patient and cause severe problems in the short and long term.

- <u>Sensory input</u> such as excessive (or even low level) noise, light and even vibration or a sense of movement (as felt when travelling by car or ambulance) can be unbearable and extremely painful for the severe M.E. patients and cause significant problems. The problem here is not merely pain in the ears and painful or burning eyes. Even low levels of noise or light (and other sensory input) can cause a significant and prolonged worsening of the severity of the condition overall, as well as symptoms including seizures, mental confusion and inability to process even very simple information, episodes of paralysis, problems with proprioception, balance and so on. Pain levels can quickly soar to a 10/10 level even with moderate or brief noise or light exposure, and recovery can be prolonged. Travelling by car is excruciating with severe M.E. and can cause a prolonged, *or permanent*, worsening of neurological, cardiac and other problems. It can also cause death (see section/question 1 below).

Note too that travelling by car causes relapse even if light, noise and vibration are minimised as much as possible. The problem isn't *just* excess sensory input. Even then, as one M.E. patient explains it, it is also the exertion of <u>movement through space</u> that leaves severe M.E. patients 'in a coma-like state' and feeling as if they're 'going into total organ failure' during and after travelling.

- A patient's <u>inability to be upright</u> for any amount of time can be very severe. Often trips out of the house, even where a patient is transferred by bed almost entirely, still require a patient to sit up for short periods which can be unbearable for the very ill and cause significant problems in the short and long term. Even sitting up in bed propped up by a few pillows counts as 'being upright' when someone is severely affected, and even a few minutes of being upright may be long enough to cause major problems.

- <u>Exposure to warm or cool temperatures</u> can also cause sometimes acute problems (as patients with M.E. have a loss of thermoregulation).

- <u>Exposure to chemicals</u> in new environments (from common personal care products worn by others, to chemicals used in building or cleaning) can cause pain, headaches and other symptoms in some patients, as can <u>exposure to mouldy environments</u>. An M.E. sufferer may be adversely affected by a level of chemicals or mould which is not detectable, or only barely detectable, by a healthy person. Not every M.E. patients is affected significantly by chemical and mould exposures but for some this is a significant problem.

- Patients with M.E. often also have very <u>restricted diets</u> (due to digestion problems, food allergies and intolerances etc.), and problems with going for even a few hours, or more than half an hour in some cases, without food (as with other patients with severe metabolic/mitochondrial disorders). There is also a need to have continual access to adequate <u>water</u>. Trips out of the house that don't accommodate these needs can make the patient very ill.

So as you can see, merely protecting the patient from physical overexertion is not enough by itself to make an activity safe for an M.E. patient. It is more complicated than that unfortunately.

One of the main misconceptions is that while walking a few steps requires additional bodily resources and cardiac output, time spent thinking, looking, listening or experiencing other sensory stimuli does not. This is not the case. Not only physical effort, but also cognitive effort, requires additional resources which an M.E. patient may not have. The brain contains some 100 billion neurons connected to 10,000 relay stations and this enormous electrical activity creates a massive need for energy and other bodily resources. The brain uses up to 25% of the entire body's demand for glucose, 25% of the blood pumped from the heart goes to the brain and the brain also needs 25% of the body's oxygen supply. (Blood supplies nutrients like glucose, protein, trace elements, and oxygen to the brain.) So of course, every extra second of 'electrical activity' – every thought, every feeling, every noise heard or sight seen – requires additional cardiac output, makes additional oxygen and glucose demands, and so on, in just the same way as does a physical activity such as walking; if not more so. So in addition to physical activity, the list of things that can cause similar severe relapse in M.E. patients also includes cognitive exertion, sensory input and orthostatic stress; anything that makes the body work harder or have to adjust in some way (Dowsett n.d. d, [Online]).

Again, that is why hospital trips (or any travelling out of the house) should be an absolute last resort for patients with severe M.E. and should be avoided wherever possible. It is counter-productive and cruel.

People with severe M.E. are some of the most vulnerable members of society and they deserve and desperately need appropriate care; care given in the home as much as possible.

It is unreasonable that these already very ill patients have to be made so much more ill to get the basic care they need, most of which could easily be administered at home at an immensely reduced physical cost to the patient.

Advice for carers

If there is a genuine need for a trip out of the house there are things that can and must be done to help minimise the harm caused. So what are the top 10 most obvious things that need to be considered by anyone providing care to an M.E. patient on a daily basis, whether at home, in transit or during a short trip to hospital?

1. Reduce exposure to light
2. Reduce exposure to noise
3. Reduce/eliminate all non-essential visitors
4. Do not encourage patients to be more physically active (or upright longer) than they can easily tolerate
5. Try to schedule demanding tasks for the patient's best time of day as much as is possible
6. Try to reduce the patient's levels of cognitive exertion and sensory input
7. Be aware of any special dietary requirements
8. Be aware of the likelihood of negative drug reactions
9. Be aware of problems with sleep and the need for extensive rest
10. Be aware that these aforementioned relapses can be delayed, and that they can be very serious and prolonged

Each of these points is expanded upon in the text: Hospital or carer notes for M.E. available on the HFME website or in the book 'Caring for the M.E. Patient.'

It's a lot to take in all at once, but everything that you can do to reduce the relapse from a hospital stay – or even better, avoid a hospital stay completely – will make a real difference and be much appreciated. Just do your best.

There is a huge difference between a 2 month long relapse and a 6 month relapse; between symptoms worsening during this time to a 7/10 level rather than a 10/10 level; between a short-term and a permanent worsening of symptoms.

(M.E. patients appreciate what a hassle it is to accommodate the demands of M.E. only too well. Those of us who have M.E. went from being normal and healthy one day to having to cope with great limits and disabilities the next, even from one hour to the next. M.E. patients understand that M.E. is a huge hassle to deal with on just about every level; we understand the issues carers grapple with.)

Conclusion

Some tasks are physically impossible for some M.E. sufferers, and others are possible but unwise. Sometimes difficult tasks can be done so long as it is in a controlled way; and strictly limited as to frequency and/or duration. Another way to say this is that some activities need to be very carefully 'rationed.' In addition, some tasks are only possible at the patient's best time of day, or with a period of rest beforehand (lasting minutes, hours or days or longer) or can only be completed if the task is modified in some way, or with assistance from a carer.

Activities that would be trivial for healthy people – including being out of bed or leaving the house for brief periods – can have disastrous consequences for patients with severe M.E. Consequences can include extremely severe and prolonged relapses, additional disability, permanent bodily damage and death.

Again, it is vitally important that M.E. patients are allowed to judge *for themselves* how much activity it is safe and wise for them to attempt. Patients are the best judges of their own limits, and patients' judgements must not be over-ruled. Patients should never be advised, encouraged or forced to be more active than their severely damaged bodies can handle; these decisions cannot safely or ethically be made by any third party.

If a patient says they *cannot* or *should not* do something: then family, friends, doctors, carers and hospital staff *must listen.* Thank you for taking the time to read this paper.

Notes on this text

1. What does 'rest' mean exactly in this context?

Resting means completely different things at different severity levels of illness. For the mildly ill resting may mean watching TV or perhaps sitting in a chair reading a book or having a quiet night in with friends. For the severely ill, these activities are not at all restful and indeed would provoke relapses. For the very severely ill, resting means lying down in a dark room, in silence and with no sensory input at all (such as TV or radio or light) and not moving physically or engaging in any type of cognitive activity. For the very severely ill a better term would be 'complete incapacitation,' rather than 'resting' as the level of inactivity is not optional. For moderately ill patients resting means something somewhere between the two extremes.

Of course for the very severely ill there will be no safe or symptom-free activity limit. Concepts of pacing or of keeping activity at a level which does not cause immediate or delayed symptoms are useless. Indeed, a sizeable proportion of the very severely ill may well be so badly affected in the first place BECAUSE of overexertion in the early stages of their illness - because they were not told how important it was to rest or were not allowed to rest adequately. This is extremely common in M.E.

This is not about patients being as inactive as possible. Of course a person with moderate M.E. does not need to live with the same restrictions as someone with severe M.E. The point here is just that patients must stay within their individual post-illness limits. Increasing the activity levels of someone with M.E. beyond their individual limits can only ever be harmful. It really doesn't matter if this is done gradually or all at once.

The evidence which shows that some 'CFS' patients are merely deconditioned and can be restored to health through graded exercise programs is based on patients who DO NOT have M.E. It should go without saying that treatment of one disease cannot be determined by studying a completely different and unrelated (and mixed) patient group. Yet this essential medical and logical guideline is all too often ignored when it comes to M.E. For more see: Smoke and Mirrors on the website. To summarise:

- No one with M.E. is *too* restrictive with their activity levels and M.E. patients do not underestimate their activity levels
- It is very difficult for M.E. patients to restrict their activity levels, and requires a high level of discipline
- M.E. patients know from bitter experience the negative consequences of overexertion
- The appropriate activity level depends on the severity of each patient's illness
- The symptoms of M.E. are not caused by deconditioning. Graded exercise does not help M.E.; if a patient improves with graded activity, they do not suffer from M.E.
- Some patients that qualify for a 'CFS' diagnosis may improve with graded exercise, but these patients do not suffer from M.E.

Acknowledgments
Thanks to Lesley Ben and Emma Searle for editing this paper. Thanks also to everyone who offered suggestions and comments as I was writing this paper.

Relevant quotes
'Those who are most injured or die are easily recognized at disease onset or shortly after as CNS, cardiovascular, or organ injury. Because of their overwhelming illness and the specificity of the end-organ injury, they are never diagnosed as M.E. except in epidemic or cluster situations.'
DR BYRON HYDE

'Documented deaths have occurred in several M.E. epidemics, but are best documented in the Cumberland epidemic and were well known in the Akureyri epidemic. All of these deaths involved CNS injury.'
DR BYRON HYDE

'Following successful immunisation against poliomyelitis in the early 1960s and the removal of 3 strains of polio virus from general circulation in the countries concerned, the related non-polio enteroviruses rapidly

filled the vacancy. By 1961, the prevalence of diseases (such as viral meningitis) caused by these agents soared to new heights. In the mid 1980s, the incidence of M.E. had increased by some seven times in Canada and the UK, while in the USA a major outbreak at Lake Tahoe (wrongly ascribed at first to a herpes virus) led to calls for a new name and new definition for the disease, more descriptive of herpes infection. This definition based on "fatigue" (a symptom common to hundreds of diseases and to normal life, but not a distinguishing feature of Myalgic Encephalomyelitis) was designed to facilitate research funded by the manufacturers of new anti-herpes drugs.'
DR ELIZABETH DOWSETT

'Under epidemic and primary M.E. there is no consensus as to the viral or infectious cause. Much of this lack of consensus may be due to the non-separation of acute onset from gradual onset patients in the M.E. and CFS groups of patients. Primary M.E. is always an acute onset illness.

Doctors A. Gilliam, A. Melvin Ramsay and Elizabeth Dowsett, John Richardson, W.H. Lyle, Elizabeth Bell of Ruckhill Hospital, James Mowbray of St Mary's and Peter Behan all believed that the majority of primary M.E. patients fell ill following exposure to an enterovirus (Poliovirus, ECHO, Coxsackie and the numbered viruses are the significant viruses in this group).

I share this belief. Recent publications have also identified the fact that enteroviruses are one of the most likely causes of M.E. If this belief is correct, many if not most of the M.E. illnesses could be vanquished by simply adding essential enteroviral genetic material from these enteroviruses to complement polio immunization.'
THE NIGHTINGALE DEFINITION OF M.E. BY DR BYRON HYDE 2006

'The failure to agree on firm diagnostic criteria has distorted the data base for epidemiological and other research, thus denying recognition of the unique epidemiological pattern of myalgic encephalomyelitis.'
DR A. MELVIN RAMSAY

'We are indebted to Dr Ramsay, an outstanding infectious disease specialist who devoted much effort to the investigation of our disease from the time that he was confronted with the epidemic at the London hospitals in the 1950's. Dr. Ramsay's fame and standing are no accident and we can see that his descriptions of what make this disease unique are accurate and Ramsay's M.E. is the same disease we have today. It is clear that attempts at confusion and name changes would serve to obscure its history and also its origins. So we must never forget Ramsay. The worldwide epidemic we have today is the same disease that Ramsay encountered many years ago.'
THE COMMITTEE FOR JUSTICE AND RECOGNITION FOR M.E.

'Despite the claims of some psychiatrists, it is not true that there is no evidence of inflammation of the brain and spinal cord in M.E.; there is, but these psychiatrists ignore or deny that evidence. It is true that there is no evidence of inflammation of the brain or spinal cord in states of chronic fatigue or 'tiredness.''
THE TERMINOLOGY OF M.E. & CFS BY PROFESSOR MALCOLM HOOPER

ON THE LACK OF FUNDING GIVEN TO LEGITIMATE M.E. RESEARCH, DR BYRON HYDE WRITES: 'Without heed, we are sitting on the edge of a cliff, waiting for disaster. For many sufferers of M.E. that disaster is already here, and few are listening.'

'Myalgic Encephalomyelitis is (a) an acute onset, (b) biphasic (c) epidemic and endemic (sporadic) contagious illness, with a (d) minimal incubation period of 4-5 days that (e) spreads easily to both children and adults. In most M.E. epidemics the affected age group is over 20, so it is obvious that this is a rapidly mutating virus or the patient has not previously contracted this virus and has developed no previous immunity. Since 1934, M.E. has been associated with over 60 epidemics worldwide. In M.E. there is always a persistent diffuse vascular injury of the CNS measurable in the acute and chronic phase. Chronic M.E. affects the body's metabolic and control mechanisms.'
DR BYRON HYDE 2011

'There is a principle which is a bar against all information, which is proof against all argument, and which cannot fail to keep man in everlasting ignorance. That principle is condemnation without investigation.'
WILLIAM PALEY (1743-1805)

CHAPTER FOUR
HFME information, references and final comments

This chapter includes the following papers and sections:

1. How, when and why HFME was founded

2. The aims of HFME and the reasons for the aims of HFME

3. Why hummingbirds as a metaphor for M.E.?

4. The HFME reference list

5. Additional HFME resources available online

6. Afterword

How, when and why HFME was founded

 HFME was founded in May 2009. The leader and founder of the group is Jodi Bassett. Jodi Bassett is an Australian writer, artist, graphic designer, and patient advocate.

Jodi contracted Myalgic Encephalomyelitis (M.E.) in 1995 when she was just 19. She went from being healthy and happy one day, to very ill and disabled with the neurological disease M.E. the next. When first ill, Jodi was reduced to 40% of her pre-illness activity level. Due to inappropriate medical advice leading to sustained overexertion (which causes serious and permanent bodily damage in M.E.), Jodi's illness quickly went from moderate to extremely severe. By 1999 she was capable of less than even 5% of her pre-illness activity level.

After more than a decade of the disease becoming worse as each year passed, Jodi's disability level finally began to stabilise. Thanks to appropriate care, education and support, her condition improved from *extremely severe* to *severe* in 2007. At the time of writing her condition continues, with careful management, to improve very slowly month by month.

She still requires the help of part-time carers to live, and is currently severely affected, housebound and largely bedbound. All of her activism and advocacy has been conducted from her bed using a laptop and a reclining laptop stand. Jodi is able to spend just 30 to 45 minutes a day (on average) on M.E. advocacy.

In 2004, Jodi Bassett started the 'A Hummingbirds' Guide to M.E.' website to try to improve awareness of the facts of M.E., and to stop other M.E. patients from being needlessly made far more ill and disabled due to inappropriate medical advice based on the false notion that M.E. is the same thing as 'CFS.' In 2009, with the help of a group of similarly-minded M.E. advocates from around the world, Jodi founded 'The Hummingbirds' Foundation for M.E.' in order to advocate for M.E. patients on a much bigger scale and to get the relevant information to a much wider audience worldwide.

For the same or similar reasons, the majority of HFME contributors are likewise disabled. There is very little advocacy for M.E. patients, and HFME contributors have determined that despite their high disability levels, they must do what they can for M.E. advocacy. The vast majority of charities that started out advocating for M.E. patients are now actively supporting the same misinformation they were created to oppose. This is helped immeasurably by the bogus concept of 'ME/CFS.' For 20 years now, M.E. patients have been subjected to serious medical neglect and abuse, even unto death in some cases. The situation is actually worsening as slick, faux advocacy groups gain more and more popularity and support from uneducated and misinformed – and often misdiagnosed – patients.

HFME is run by and for M.E. patients. HFME contributors also aim to advocate for those non-M.E. patients given the always meaningless 'CFS' diagnosis who also are not being served well by the various 'CFS' charities, and who are also harmed by the bogus disease category of 'CFS' and the overwhelming triumph of financial greed over ethics, science and basic human rights.

'The greater danger for most of us is not that our aim is too high and we miss it, but that it is too low and we reach it.'
MICHELANGELO BUONARROTI

The aims of HFME

 **Aim 1:** To disseminate scientifically accurate information on Myalgic Encephalomyelitis (M.E.) to M.E. patients; to their carers, family and friends; to the medical profession and other professions which deal with M.E. patients; to policy makers; to M.E. advocates and activists and to the general public, as per the paper What is M.E.? and as further discussed on the HFME website and in the HFME books.

Reason for Aim 1: An important fact about M.E. is that it can be made very much worse by overexertion. Overexertion can also cause death in M.E. It is vital that M.E. patients should be aware of the importance of avoiding overexertion in order to avoid needless permanent bodily damage. Many M.E. patients have become severely affected, bed-bound, or in constant pain because of overexertion and some patients have died. Tragically, these negative outcomes happen all too often, due to inappropriate medical advice.

M.E. patients currently receive little or no helpful medical advice, as most doctors have little understanding of the disease. Therefore it is essential that advice from M.E. experts on treatment and management of the disease should be made available to patients.

Aim 2: To oppose false and meaningless disease categories such as 'CFS,' 'CFIDS,' 'ME/CFS,' 'CFS/ME,' 'ME-CFS' and Myalgic 'Encephalopathy,' as per the papers What is M.E.? and M.E. is not fatigue, or 'CFS' and as further discussed on the HFME website and in the HFME books. These bogus disease categories and concepts must be abandoned for the benefit of all the different patient groups involved.

Reason for Aim 2: The fact that a person receives a diagnosis of 'CFS' (a) does not mean that the patient has M.E., and (b) does not mean that the patient has any other distinct and specific illness named 'CFS.' 'CFS' can only ever be a misdiagnosis, and prevents patients from getting a correct diagnosis and appropriate (even curative or life-saving) treatment. See: The misdiagnosis of CFS.

Currently few doctors or researchers recognise M.E. Patients receive little or no helpful medical advice, as most doctors have an entirely inaccurate understanding of the disease based on confusion with the bogus disease category of 'CFS.'

Concepts such as 'ME/CFS,' 'CFS/ME' and 'ME-CFS' are just as problematic and meaningless as 'CFS' and in many ways more so, as they incorrectly imply that M.E. and 'CFS' are synonymous terms. The only groups which gain from the continuation of these fictional disease categories (to the detriment of patients) are vested interest groups, such as:

1. Medical insurance companies
2. Governments
3. The vaccine industry
4. The chemical industry
5. Psychiatrists
6. 'CFS specialists'
7. Medical doctors
8. The media (including medical journals)
9. 'CFS' or 'ME/CFS' (and other) groups that sell vitamins and other supplements to 'CFS' patients
10. 'CFS' or 'ME/CFS' so-called patient support and advocacy groups

For more information see: Who benefits from 'CFS' and ME/CFS'?

Aim 3: To broaden the online and offline presence of HFME in order to disseminate information about M.E. and to correct misinformation about 'CFS' as per aims 1 and 2 above. This will involve improving internet accessibility as well as raising the profile of the website so that it can be found easily by M.E. patients as well as those misdiagnosed with 'CFS' who have other diseases. The information on the HFME website will also be made available in a convenient book format. Several book releases are planned.

Reason for Aim 3: There are currently very few websites or books available which accurately describe the historical, political and medical facts of M.E. and of 'CFS' and which also offer research papers by the world's leading M.E. experts, papers on various aspects of the disease (which are vital reading for patients and carers) and factual information and support. The HFME site and books aim to fill this gap.

For every factual website there are at least a hundred websites which support the misinformation about M.E. and 'CFS' which causes patients so much harm. The same is true about books on this subject.

M.E. patients suffer greatly from a lack of information on their condition, and urgently need the information offered by the HFME website and books, as do those patients misdiagnosed with 'CFS' who have other diseases. This information also urgently needs to become readily available to doctors, carers, human rights groups, the media and politicians.

Aim 4: To make it clear that M.E. is not 'medically unexplained' or 'mysterious' as 'CFS' is and that an abundance of scientific evidence already exists which proves that M.E. is a disabling and potentially fatal neurological disease. HFME makes available this valuable research which is generally overlooked.

Reason for Aim 4: The historical facts of M.E. and the available scientific research on M.E. is being **actively suppressed** and **deliberately ignored** due to vested political interests.

The scientific research on M.E. is generally overlooked in favour of misinformation promoted by vested interest groups incorporating fatigued patients who have been given a misdiagnosis of 'CFS.'

Aim 5: To defend the M.E. community (and those with non-M.E. diseases misdiagnosed as 'CFS') against counter-productive 'activism' strategies such as renaming 'CFS' with some variation of the term M.E.

Reason for Aim 5: Such terms and concepts obscure the reality of M.E., equate it with 'CFS' and harm the cause of M.E. These unscientific terms and concepts also cause harm to those with non-M.E. diseases misdiagnosed as 'CFS.'

Aim 6: To promote appropriate research based on a proper understanding of M.E., and to oppose flawed concepts such as the 'subgroups' of 'CFS' or 'ME/CFS' concept.

Reason for Aim 6: Studying subgroups of heterogeneous groups of fatigue patients does not in any way help M.E. patients, or any other distinct patient group.

Aim 7: To be a voice for those suffering from M.E. who are facing mistreatment and abuse due to the false notion that M.E. is the same thing as 'CFS' and is a trivial and 'mysterious' illness or a mental illness characterised by 'fatigue.'

Reason for Aim 7: Many patients with M.E. are subjected to medical abuse, mistreatment from social services, lack of understanding from the general public and even ridicule, neglect and abuse from friends and family. M.E. patients are (with very few exceptions) **not being served** by the charities which are supposed to represent them. The vast majority of charities purporting to help M.E. patients worldwide, whatever their original aims, are now actively supporting the propaganda which harms M.E. patients.

Aim 8: To be a voice for all those patients misdiagnosed with 'CFS' who do not have M.E., but other illnesses including: cancer, fibromyalgia, various post-viral fatigue syndromes, athlete's over-training syndrome, Lyme disease, Behcet's disease, PTSD, depression and other mental illnesses, burnout, thyroid or adrenal diseases, various vitamin-deficiency diseases, and so on. To encourage each of these patients to reject their 'CFS' misdiagnosis and seek a correct diagnosis and appropriate treatment.

Reason for Aim 8: A diagnosis of 'CFS' can only ever be a misdiagnosis. Currently, many hundreds of thousands of patients with a vast array of different diseases are misdiagnosed with 'CFS' and this can cause needless suffering and disability, sometimes leading to needless deaths.

There are many 'CFS' advocacy groups out there, but most tell patients that their 'CFS is real' and that 'CFS is not a mental illness' and that they must join the fight to legitimise 'CFS' and so on. This may placate patients in the short term, and make them feel as if they are being helped, but it misses the point entirely and is counter-productive. This approach keeps patients ignorant of the basic information that can improve their situation and their medical outcome.

Defending 'CFS' only aids vested interest groups; patients will benefit when 'CFS' is abandoned. Every 'CFS' misdiagnosed patient deserves to know that 'CFS' does not exist and that it should never be considered the end point of diagnosis.

Some of the conditions commonly misdiagnosed as 'CFS' are well defined and well-known illnesses that are treatable, but only once they have been correctly diagnosed. Some conditions are also very serious or can even be fatal if not correctly diagnosed and managed. Every patient deserves the best possible opportunity for appropriate treatment for their illness, and for recovery. This process must begin with a correct diagnosis. A correct diagnosis is half the battle won.

Aim 9: To enlist the help of human rights groups, medical professionals and the quality media to help to achieve the above stated goals as is their obligation and duty, a duty that has unfortunately been almost completely ignored for the last 20 years (with a few notable exceptions).

Aim 10: To obtain funding to pursue the aims described previously.

Quotes

'To the very few physicians still practicing today who began seeing patients with this illness some 40 years ago and who have continued to record and publish their clinical findings throughout, the current enthusiasm for renaming and reassigning this serious disability to subgroups of putative and vague "fatigue" entities, must appear more of a marketing exercise than a rational basis for essential international research. It was not always so unnecessarily complicated!'
REDEFINITIONS OF M.E. - A 20TH CENTURY PHENOMENON BY DR ELIZABETH DOWSETT

'The world is a dangerous place to live; not because of the people who are evil, but because of the people who don't do anything about it.'
ALBERT EINSTEIN

'I must admit that I personally measure success in terms of the contributions an individual makes to her or his fellow human beings.'
MARGARET MEAD

'Never lose an opportunity of urging a practical beginning, however small, for it is wonderful how often in such matters the mustard-seed germinates and roots itself.'
FLORENCE NIGHTINGALE

Hummingbirds

Some time ago I was flicking through a book (looking for some artistic inspiration) when I came upon a stage-by-stage illustration of hummingbirds hovering and it struck a chord in me. Soon it hit me why. In the same way a hummingbird comes crashing to the ground with a big SPLAT! if it falters in the complex series of movements that keep it in the air, in a different sort of way, so do I.

I contracted Myalgic Encephalomyelitis (M.E.) in 1995 when I was 19. One day I was healthy and the next day absolutely everything changed. Since then I've been forced to keep on 'flapping my wings' endlessly lest I fall into an even more agony-filled and semi-conscious paralysed heap. I have to constantly remain aware of, and quickly adjust to, all sorts of small changes in my environment and my body. My version goes something like this;

FLAP! Making sure I don't spend too much time flat in bed (consecutively), or my vertigo becomes much more severe, the room spins horribly and I feel I am falling over backwards as I try to walk, or have to struggle not to fall off the edge of my perfectly flat bed.

FLAP! Trying not to stand or sit up for too long or my heart just can't cope and it struggles to beat properly and I feel extremely ill for hours afterward. It feels like a heart attack in every organ. Tests show my heart-rate can climb as high as 170 beats-per-minute just from a few minutes of 'exertion.'

Then I forget for just a few moments about having to be careful about how much light I expose my eyes to and instantly...C
R
A
S
H!

Burning pain that lasts for hours leaving me unable to open my eyes. But still I can't let my guard down and have to get myself back in the air straight away…

FLAP, FLAP, FLAP! I manage to quickly close all my doors and put my headphones on to block out some neighbourhood noise that would have left me in agonising pain, experiencing seizures, memory loss and taking five days to recover from if I'd listened to it at full volume.

Then I forget to avoid one of the foods I am intolerant of (but that I tolerated perfectly well the day before) and a few minutes later...**T H U D!** Abdominal pain, headache, bloating, severe itching and nausea for hours afterward. But quickly I have to get myself back up in the air…

FLAP! I manage to make my bath neither too cold (which leaves me shaking and unable to get warm for hours), or too hot (which makes me light-headed, my heartbeat becoming irregular and fluttery for the next six hours so it feels like my heart is struggling to beat and I'm having a heart attack.)

Then I forget to put my blanket over myself properly and within a short while...**S P L A T!** I get so cold I can't get myself warm again and it turns into a horrible shivering fever, which leads to delirium, paralysis and eventually loss of consciousness for several hours. I then spend the rest of the day partly paralysed and feeling (neurologically and cognitively) as if I'd had a stroke.

Because so many normal everyday things cause me to 'crash' I have to constantly monitor every little thing I do and every aspect of my environment to try and keep myself 'in the air' as much as possible. It's a never-ending task and a fairly thankless one too, as my highest level of functioning is pretty low anyway – I'm 100% house-bound and 99% bed-bound on my <u>best</u> days, I'm in continual pain and experience many different neurological, cardiac, cognitive and other symptoms constantly. But it's not so much having a painful and limited life that is so hard to bear (though obviously that's part of it), but to have to plan and work so endlessly hard every minute just to keep my life this 'good': that's what *really* makes it a nightmare.

With a bit more research however, I quickly found a more positive reason to identify with hummingbirds. You see, although at first glance they are tiny, seemingly defenceless and extremely vulnerable to attack from anyone or anything, they are actually quite tough little critters. They never back down from a fight even if the odds are against them, taking on other birds much larger than they are when they need to. What their bodies lack in strength and power is made up for by their bravery, strength of mind and *spirit*.

I've met so many people with M.E. that share that same spirit, particularly with severe M.E.; people that have remained kind, witty, giving, optimistic and determined to make the best of what they have despite dealing with an unbelievably severe (and potentially fatal) neurological, cardiac and metabolic disease often without the support of family, friends or the health and welfare systems; indeed, often with direct opposition, criticism and sometimes abuse from these people and organisations.

I consider these people no less beautiful inside than a hummingbird is to the eye. The human spirit is capable of amazing resilience and endurance and I can see no greater example of this than people suffering from severe M.E. I think they are truly inspiring. When every hour of every day is so difficult and there's no foreseeable end in sight, the fact that along with the obvious sadness and frustration there can also be hope and humour is just amazing.

The world is full of inspiring stories about people triumphing over quite small problems (compared to M.E.), always with the support of everyone around them and much back patting and praise when they've finished their short 'ordeal'. There is nothing wrong with that, except that on the other hand, there are desperately ill people with severe M.E. who have no support at all yet are able to somehow keep going through one horrendous ordeal of a day after another, often for many years or even decades. Not only are they rarely acknowledged for their hard work and amazing strength, but they are sometimes actually labelled as malingerers, or seen as mentally weak or defective in some way. Sometimes they are, unbelievably, treated as if they were merely very 'fatigued' or 'tired all the time' instead of very ill with a severe neurological and cardiovascular disease. It really does boggle the mind that there can be such a gap between perception and reality.

I've since featured hummingbirds in many of my paintings and drawings and this is why. I see the same sort of strength and beauty, combined with such heartbreaking vulnerability in my M.E. friends everyday. Nothing I've seen on this earth is more inspiring to me, more beautiful, *or* more tragic, more heartbreaking. **I think of people with severe M.E. as hummingbirds now** – vulnerable, strong and strikingly beautiful all at once; and more than overdue for some consideration, compassion and care in this world.

This paper is dedicated to all the beautiful 'hummingbird' M.E. patients I've met. To view more of my artwork featuring hummingbirds please visit the www.ahummingbirdsguide.com website.

Acknowledgments
Thanks to Caroline Gilliford and Emma Searle for editing this paper.

The HFME reference list

All of the information concerning Myalgic Encephalomyelitis (M.E.) in this book and on the HFME website is fully referenced and has been compiled using the highest quality resources available, produced by the world's leading M.E. experts.

More experienced and more knowledgeable M.E. experts than these – Dr Byron Hyde and Dr Elizabeth Dowsett in particular – do not exist.

Between Drs Byron Hyde and Elizabeth Dowsett, and their mentors the late Drs John Richardson and Melvin Ramsay (respectively), these four doctors have been involved with M.E. research and M.E. patients for well over 100 years collectively, from the 1950s to the present day. Between them they have examined more than 15,000 individual (sporadic and epidemic) M.E. patients, as well as each authoring numerous studies and articles on M.E., and books (or chapters in books) on M.E. Again, more experienced, more knowledgeable and more credible M.E. experts than these simply do not exist.

The chapters in this book to a large extent provide merely a summary of the most important facts about M.E. They have been written for the benefit of those individuals without the time, inclination or ability to read each of the large number of far more complex and lengthy source papers. A full list of references is given at the end of most papers however, and these original references are highly recommended as essential additional reading for any physician (or anyone else) with a real interest in M.E.

Governments around the world are currently spending $0 a year on M.E. research. Considering the severity of the illness, and the vast numbers of patients involved, ranging in age from small children to adults, this is a worldwide disgrace. The fiction of 'CFS' represents outright medical fraud, involving serious medical abuse and neglect of patients, on a massive scale.

Not everyone was taken in by the 'CFS' insurance scam, however. A small but dedicated group of M.E. experts have each examined many thousands of M.E. patients and have made many remarkable discoveries about the pathology of M.E. These discoveries have confirmed many times over what was already known about M.E. prior to 1988, before M.E. research became tainted by irrelevant concepts of 'fatigue' and 'CFS' and then disappeared almost entirely.

Before reading the information in the reference list, please be aware of the following facts:

1. Myalgic Encephalomyelitis and 'Chronic Fatigue Syndrome' are not synonymous terms. The overwhelming majority of research on 'CFS' or 'CFIDS' or 'ME/CFS' or 'CFS/ME' or 'ICD-CFS' does not involve M.E. patients and is not relevant *in any way* to M.E. patients. However, if the M.E. community were to reject all 'CFS' labelled research as *only* relating to 'CFS' patients (including research which describes those abnormalities and characteristics unique to M.E. patients), this may support the myth that 'CFS' is just a 'watered down' definition of M.E. and that M.E. and 'CFS' are virtually the same thing and share many characteristics.

A very small number of 'CFS' studies refer in part to people with M.E. but it may not always be clear which parts refer to M.E. The A warning on 'CFS' and 'ME/CFS' research and advocacy paper is recommended reading and includes a checklist to help readers assess the relevance of individual 'CFS' studies to M.E. (if any) and explains some of the problems with this heterogeneous and skewed research.

In future, it is essential that M.E. research again be conducted using only M.E. defined patients and using only the term M.E. The bogus, financially-motivated disease category of 'CFS' must be abandoned.

2. The research referred to in the list below varies considerably in quality. Some is of a high scientific standard and relates wholly to M.E. and uses the correct terminology. Other studies or articles are included which may only have partial or minor possible relevance to M.E., use unscientific terms/concepts such as 'CFS,' 'ME/CFS,' 'CFS/ME,' 'CFIDS' or Myalgic 'Encephalopathy' and also include a significant amount of misinformation. Before reading this research it is also essential that the reader be aware of the most commonly used 'CFS' propaganda, as explained in A warning on 'CFS' and 'ME/CFS' research and advocacy and in more detail in Putting research and articles on M.E. into context.

Partial HFME reference list (see www.hfme.org for a full list):

1. Acheson, AD 1954, *Encephalomyelitis associated with poliomyelitis virus,* Lancet: Nov 20th 1954:1044-1048
2. Acheson, AD 1956, *A New Clinical Entity?* THE LANCET. LONDON : May 26 1956:789-790
3. Acheson, AD 1955, *Outbreak at The Royal Free,* Lancet 20 August 1955:304-305
4. Acheson, AD 1959, *The Clinical Syndrome Variously Called Benign Myalgic Encephalomyelitis, Iceland Disease and Epidemic Neuromyasthenia,* Am J Med 1959:569 595 (Also: In *The Clinical and Scientific Basis of Myalgic Encephalomyelitis,* Hyde, Byron M.D. (ed) The Nightingale Foundation, Ottawa, pp. 129-158.)
5. Bassett, Jodi 2010, *The Ultra-Comprehensive Myalgic Encephalomyelitis Symptom List* [Online], Available: http://www.hfme.org/themesymptomlist.htm
6. Bassett, Jodi 2010, *What is Myalgic Encephalomyelitis* [Online], Available: http://www.hfme.org/whatisme.htm
7. Bastien, Sheila PhD. 1992, *Patterns of Neuropsychological Abnormalities and Cognitive Impairment in Adults and Children* in Hyde, Byron M.D. (ed) 1992, *The Clinical and Scientific Basis of Myalgic Encephalomyelitis,* Nightingale Research Foundation, Ottawa
8. Dchan W, Gow JW & Curtis F 1999, *Blood Brain Barrier Breakdown in Myalgic Encephalomyelitis,* Presented at "Fatigue 2000" Conference, London 23rd 24th April 1999, arranged by The National ME Centre, Harold Wood, Essex, in conjunction with the Essex Neurosciences Unit
9. Bell, EJ & McCartney, RA. 1984, A study of Coxsackie B virus infections 1972-1983, J Hyg (Lond). 1984 Oct;93(2):197-203.
10. Bell EJ, McCartney RA & Riding MH 1988, Coxsackie B viruses and myalgic encephalomyelitis, Ruchill Hospital, Glasgow, J R Soc Med. 1988 Jun;81(6):329-31.
11. Bell, David S MD 1995, *The Doctor's Guide to CFIDS,* Perseus Books, Massachusetts
12. Bodian D, 1949, *Histopathological basis of findings in poliomyelitis,* American Journal of Medicine, 1949;6:563-578
13. Bruno, RL, Frick NM, & Creange SJ et al. 1997, *A brain model for post viral fatigue syndrome,* ME Today, 1997;5/6:18-21
14. Bruno RL. et al. 1998, *Parallel between Post-Polio Fatigue and Chronic Fatigue Syndrome – A common Pathophysiology?,* American Journal of Medicine. 1998 105 (3A) : 66(s) – 73(s)
15. Bruno, RL 2002, *THE POLIO PARADOX,* Chapter 11: 164-166. Warner Books INC 2002, 1271 Avenue of the Americas, NY 10020.
16. Chaburksy, Borys. Hyde, Byron M.D. & Anil Jain M.D. 1992, *A Description of Patients who Present with a Presumed Diagnosis of M.E.* in Hyde, Byron M.D. (ed) 1992, *The Clinical and Scientific Basis of Myalgic Encephalomyelitis,* Nightingale Research Foundation, Ottawa, pp. 19-24
17. Cheney, Paul M.D. 2006, *The Heart of the Matter* [video recording], Available: http://www.hfme.org/wcheney.htm
18. Chia, John kai-sheng & Chia, Andrew Y, 2007, *Chronic fatigue syndrome is associated with chronic enterovirus infection of the stomach,* EV Med Research, United States, J Clin Pathol doi:10.1136/jcp.2007.050054

19. Chia JK 2005, *The role of enterovirus in chronic fatigue syndrome,* J Clin Pathol. 2005 Nov;58(11):1126-32, CEI Research Center, Torrance, CA 90505, USA.
20. Chia J, Chia A, Voeller M, Lee T, Chang R. 2010, *Acute enterovirus infection followed by myalgic encephalomyelitis/chronic fatigue syndrome (ME/CFS) and viral persistence,* J Clin Pathol. 2010 Feb;63(2):165-8. Epub 2009 Oct 14, EV Med Research, Torrance, California, USA.
21. Chia J, & Chia A. 2002, *Detection of enteroviral RNA in the peripheral blood leukocytes of patients with the chronic fatigue syndrome,* Abstract 763. In: Program and abstracts of the 40th annual meeting of the Infectious Diseases Society of America. Chicago, IL: IDSA, 2002:178.
22. Colby J 2006, *Special problems of children with myalgic encephalomyelitis/chronic fatigue syndrome and the enteroviral link,* J Clin Pathol. 2007 Feb;60(2):125-8. Epub 2006 Aug 25, Tymes Trust, Ingatestone, Essex, UK.
23. Compston, Nigel D 1978, *An outbreak of encephalomyelitis in the Royal Free Hospital Group, London, in 1955,* Postgraduate Medical Journal 1978:54:722-724.
24. Cook DIB & Natelson BH et al. 2001, *Relationship of brain MRI abnormalities and physical functional status in chronic fatigue syndrome,* Int J Neurosci 2001:107: (1-2):1-6
25. Dowsett, Elizabeth MBChB. 2002a, *The impact of persistent enteroviral infection,* [Online], Available: http://www.hfme.org/wdowsett.htm
26. Dowsett, Elizabeth MBChB. 2002b, *MEDICAL RESEARCH COUNCIL DRAFT DOCUMENT FOR PUBLIC CONSULTATION,* [Online], Available: http://www.hfme.org/wdowsett.htm
27. Dowsett, Elizabeth MBChB. 2001a, *THE LATE EFFECTS OF ME Can they be distinguished from the Post-polio syndrome?* [Online], Available: http://www.hfme.org/wdowsett.htm
28. Dowsett, Elizabeth MBChB. 2001b, *A rose by any other name* [Online], Available: http://www.hfme.org/wdowsett.htm
29. Dowsett, Elizabeth MBChB. 2000, *Mobility problems in ME* [Online], Available: http://www.hfme.org/wdowsett.htm
30. Dowsett, Elizabeth MBChB. 1999a, *Redefinitions of ME* [Online], Available: http://www.hfme.org/wdowsett.htm
31. Dowsett, Elizabeth MBChB. 1999b, *Research into ME 1988 - 1998 Too much PHILOSOPHY and too little BASIC SCIENCE!,* [Online], Available: http://www.hfme.org/wdowsett.htm
32. Dowsett EG & Richardson J 1999c, The Epidemiology of Myalgic Encephalomyelitis (ME) in the UK, Evidence submitted to the All Party Parliamentary Group of Members of Parliament, 23 Nov 1999
33. Dowsett, Elizabeth MBChB. 1998, *Can Hysteria be diagnosed with confidence ? - Conflicts in British Research,* [Online], Available: http://www.hfme.org/wdowsett.htm
34. Dowsett, EG 1998a, *Enteroviral Infections and their Sequelae,* BSAEM; RCGP 1998:1-10.
35. Dowsett EG & DM Jones DM 1998a, The Organic Basis of ME / CFS, Information and Statistics presented to the Chief Medical Officer in person at a meeting on 11th March 1998
36. Dowsett E, 1988, *Human enteroviral infections,* Journal of Hospital Infection 1988:11:103-115.

37. Dowsett, EG & Colby, J 1997, *Long Term Sickness Absence due to ME/CFS in UK Schools - an epidemiological study with medical and educational implications*, Journal of Chronic Fatigue Syndrome 1997; 3 (2): 2942

38. Dowsett, Elizabeth MBChB. n.d. a, *Differences between ME and CFS*, [Online], Available: http://www.hfme.org/wdowsett.htm

39. Dowsett, Elizabeth MBChB. n.d. b, *Time to put the exercise cure to rest*, [Online], Available: http://www.hfme.org/wdowsett.htm

40. Dowsett, Elizabeth MBChB. n.d. c, *Is Stress more than a modern buzz word?*, [Online], Available: http://www.hfme.org/wdowsett.htm

41. Dowsett, Elizabeth MBChB. n.d. d, *Brain problems in ME - is there a simple explanation?* , [Online], Available: http://www.hfme.org/wdowsett.htm

42. Dowsett, Elizabeth MBChB. in: Colby, Jane 1996, *ME: The New Plague*, Ipswitch Book Company, Ipswitch.

43. Dowsett E. & Ramsay, A.M. n.d., ' *Myalgic Encephalomyelitis: Then and Now*' The Clinical and Scientific Basis of Myalgic Encephalomyelitis, B. Hyde (ed.), The Nightingale Foundation, Ottawa, pp. 81-84.

44. Dowsett E., Ramsay A.M., McCartney A.R., & Bell E.J. 1990, ' *Myalgic Encephalomyelitis: A persistent Enteroviral Infection?*' in The Clinical and Scientific Basis of Myalgic Encephalomyelitis, B. Hyde (ed.), The Nightingale Foundation, Ottawa, pp. 285-291.

45. Dunn, Linda (in consultation with the Cross Party Group on ME) 2005,*Myalgic Encephalomyelitis. The impact on sufferers: is health policy in Scotland on the right path?* [Online], Available: (Link)

46. Fegan KG, Behan PO, & Bell EJ 1983, *Myalgic encephalomyelitis--report of an epidemic*, J R Coll Gen Pract. 1983 Jun;33(251):335-7.

47. Galbraith, DN et al.1995, *Phylogenetic analysis of short enteroviral sequences from patients with chronic fatigue syndrome*, Journal of General Virology 1995:76:1701-1707.

48. Galbraith DN, Nairn C, & Clements GB. 1995, *Phylogenetic analysis of short enteroviral sequences from patients with chronic fatigue syndrome*, J Gen Virol 1995;76:1701–7.

49. Gilliam, AG 1934, *Epidemiological study of an epidemic diagnosed as poliomyelitis occurring among the personnel of Los Angeles County General Hospital during the summer of 1934*, Public Health Bulletin, US Treasury Department No. 240, 1938

50. Grufferman, S & Komaroff, AL & Bell, DS & Peterson, DL & Daugherty, S & Bastien, S et al. 1991, Considerations in the Design of Studies of Chronic Fatigue Syndrome, Reviews of Infectious Diseases, Volume 13, Supplement 1: S1 - S140. University of Chicago Press.

51. Grufferman, S. 1992, *Epidemiologic and immunologic findings in clusters of chronic fatigue syndrome*, In The Clinical and Scientific Basis of Myalgic Encephalomyelitis, Hyde, Byron M.D. (ed) The Nightingale Foundation, Ottawa, pp. 189-195

52. Hooper, M. & Marshall E.P. & Williams M. 2007, *Corporate Collusion?*, Available: http://www.hfme.org/whooper.htm

53. Hooper, M. & Marshall E.P. 2005a, *Illustrations of Clinical Observations and International Research Findings from 1955 to 2005 that demonstrate the organic aetiology of Myalgic Encephalomyelitis* [Online]. Available: http://www.hfme.org/wmarwillhoopgibsonenqui.htm

54. Hooper, M. & Marshall E.P. 2005b, *Myalgic Encephalomyelitis: Why no accountability?* [Online], Available: http://www.hfme.org/wmarwillmewna.htm

55. Hooper, M. & Montague S 2001a, *Concerns about the forthcoming UK Chief medical officer's report on Myalgic Encephalomyelitis (ME) and Chronic Fatigue Syndrome (CFS) notably the intention to advise clinicians that only limited investigations are necessary* (The Montague/Hooper paper) [Online], Available: http://www.hfme.org/whooper.htm

56. Hooper, M. & Montague S 2001b, *Concepts of accountability* [Online], Available: http://www.hfme.org/whooper.htm

57. Hooper, M 2006, Myalgic Encephalomyelitis (ME): a review with emphasis on key findings in biomedical research J. Clin. Pathol. published online 25 Aug 2006; doi:10.1136/jcp.2006.042408 Available: http://www.hfme.org/whooper.htm

58. Hooper, M. 2003a, *The MENTAL HEALTH MOVEMENT: PERSECUTION OF PATIENTS?* [Online], Available: http://www.hfme.org/whooper.htm

59. Hooper, M 2003b, *Engaging with M.E.: Towards Understanding, Diagnosis and Treatment*, University of Sunderland, UK

60. Hooper, M. n.d., *The Terminology of ME and CFS* [Online], Available: http://www.hfme.org/whooper.htm

61. Hooper, M. Marshall E.P. & Williams, M. 2001, *What is ME? What is CFS? Information for Clinicians and Lawyers*, [Online], Available: http://www.hfme.org/wmarwillhoopwimewicfs.htm

62. Hyde, Byron M.D. 2009, *Missed Diagnoses: M.E. and CFS*, Nightingale Research Foundation, Canada

63. Hyde, Byron M.D. 2007, *The Nightingale Definition of Myalgic Encephalomyelitis* [Online], Available: www.hfme.org/whyde.htm

64. Hyde, Byron M.D. 2006, *A New and Simple Definition of Myalgic Encephalomyelitis and a New Simple Definition of Chronic Fatigue Syndrome & A Brief History of Myalgic Encephalomyelitis & An Irreverent History of Chronic Fatigue Syndrome* [Online], Available: http://www.hfme.org/whyde.htm

65. Hyde, Byron M.D. 2003, *The Complexities of Diagnosis* in (ed) Jason, Leonard at et al. 2003 *Handbook of Chronic Fatigue Syndrome* by Ross Wiley and Sons, USA. Available: http://www.hfme.org/whyde.htm

66. Hyde, Byron M.D. 1992, *Preface* in Hyde, Byron M.D. (ed) 1992, The Clinical and Scientific Basis of Myalgic Encephalomyelitis, Nightingale Research Foundation, Ottawa.

67. Hyde, B. M. 1992b, A bibliography of ME epidemics, In *The Clinical and Scientific Basis of Myalgic Encephalomyelitis*, Hyde, Byron M.D. (ed) The Nightingale Foundation, Ottawa, pp. 176-186.

68. Hyde, B. M., Biddle, R., & McNamara, T. 1992, Magnetic resonance in the diagnosis of ME/CFS, a review, In *The Clinical and Scientific Basis of Myalgic Encephalomyelitis*, Hyde, Byron M.D. (ed) The Nightingale Foundation, Ottawa, pp. 425-431.

69. Hyde, Byron M.D. & Anil Jain M.D. 1992, *Clinical Observations of Central Nervous System Dysfunction in Post Infectious, Acute Onset M.E.* in Hyde, Byron M.D. (ed) 1992, *The Clinical and Scientific Basis of Myalgic Encephalomyelitis*, Nightingale Research Foundation, Ottawa, pp. 38-65.

70. Hyde, Byron M.D. & Anil Jain M.D. 1992a, *Cardiac and Cardiovascular aspects of M.E.: A Review* in Hyde, Byron M.D. (ed) 1992, *The Clinical and Scientific Basis of Myalgic Encephalomyelitis*, Nightingale Research Foundation, Ottawa, pp. 375-383.

71. Hyde, Byron M.D., Bastien S Ph.D. & Anil Jain M.D. 1992, *General Information, Post Infectious, Acute Onset M.E.* in Hyde, Byron M.D. (ed) 1992, *The Clinical and Scientific Basis of Myalgic Encephalomyelitis*, Nightingale Research Foundation, Ottawa, pp. 25-37.

72. Hyde, Byron M.D. 1988, *Are Myalgic Encephalomyelitis and CFS Synonymous Terms?* [Online], Available: http://www.hfme.org/whyde.htm

73. Hyde, Byron MD. 2010, *Mental health problems in patients with myalgic encephalomyelitis and fibromyalgia syndrome*, [Online], Available: http://www.hfme.org/whyde.htm

74. Johnson, Hillary 1996, *Osler's Web*, Crown Publishers, New York

75. Jones, Doris M. M.S.c. 1998, *SOME FACTS AND FIGURES ON CBT, GET AND OTHER APPROACHES Directly from the 'Horses' Mouths* [Online]. Available: link

76. Keighley BD & Bell EJ, 1983, *Sporadic myalgic encephalomyelitis in a rural practice*, JRCP 1983:33:339-341.

77. Lerner AM et al, 1997, *Cardiac involvement in patients with CFS as documented with Holter monitor and biopsy data*, Infectious Diseases in Clinical Practice 1997:6:327-333

78. Lerner, MA & Corning, PD 1998, *RESEARCH BREAKTHROUGH: ME/CFS AN INFECTIOUS CARDIOMYOPATHY?* [Online], Available: http://www.hfme.org/wlerner.htm

79. Lyle, WH 1959, *An outbreak of a disease believed to have been caused by ECHO 9 Virus*, Annals of Internal Medicine 1959; 51: 248-269

80. Marshall, Eileen & Williams, Margaret. 2006a, *Some of the abnormalities that have been demonstrated in ME/CFS* [Online], Available: http://www.hfme.org/wmarwillsotathbdime.htm

81. Marshall, Eileen & Williams, Margaret. 2006b, *Inquest implications* [Online], Available: http://www.hfme.org/wmarwillinquest.htm

82. Marshall, Eileen & Williams, Margaret. 2005b, *Proof positive? Evidence of the deliberate creation via social constructionism of*

"psychosocial" illness by cult indoctrination of State agencies, and the impact of this on social and welfare policy [Online], Available: http://www.hfme.org/wmarwillpp.htm

83. Marshall, Eileen & Williams, Margaret. 2005c, *To set the record straight about Ean Proctor from the Isle of Man* [Online], Available: http://www.hfme.org/wmarwilltstrsaep.htm

84. Marshall, Eileen & Williams, Margaret. 2005d, *Profits before Patients?* [Online], Available: http://www.hfme.org/wmarwillpbp.htm

85. Martin, JW 1989, *Detection of Viral Related Sequences in CFS Patients using the Polymerase Chain Reaction*, The Nightingale Research Foundation, 1989: 1-5

86. McGarry F, Gow J and Behan PO, 1994, *Enterovirus in the chronic fatigue syndrome*, Ann Intern Med 1994:120:972 973

87. Melnick JL. Ledinko N, Kaplan A, & Kraft E. 1950, *Ohio Strains of a Virus Pathogenic for Infant Mice (Coxsackie Group). Simultaneous occurrence with poliomyelitis virus in patients with "summer grippe"*, Journal of Experimental Medicine. 1950 : 91:185-195

88. Mena, I (1991, *Study of cerebral perfusion by neuro-SPECT in patients with chronic fatigue syndrome*, Presented at Chronic Fatigue Syndrome: Current Theory and Treatment conference, Bel Air, CA.

89. Michell, Lynn 2003, *Shattered: Life with ME*, Thorsons Publishers, London

90. Mocé-Llivina L, Lucena F & Jofre J, *Enteroviruses and bacteriophages in bathing waters*, Appl Environ Microbiol. 2005 Nov;71(11):6838-44, Department of Microbiology, Faculty of Biology, University of Barcelona, Avda. Diagonal, 645 Edifici Annex, Planta 0, E-08028 Barcelona, Spain.

91. Montague, T.J., Marrie, T., Klassen, G. Bewick, D., & Horacek, B.M. 1989, *Cardiac Function at Rest and with Exercise in the Chronic Fatigue Syndrome*, April 1989, Chest, Vol 95, p779-784,.

92. Muir P et al. 1998, *Molecular typing of enteroviruses: current status and future requirements*, The European Union Concerted Action on Virus Meningitis and Encephalitis, Clin Microbiol Rev. 1998 Jan;11(1):202-27, Department of Virology, United Medical School of Guy's Hospital, London, United Kingdom.

93. Muir, P & Archard, LC 1994, *There is evidence for persistent enterovirus infection in chronic medical conditions in humans*, Reviews in Medical Virology, 1994; 4: 245-250

94. Murdoch JC 1988, *The Myalgic Encephalomyelitis Syndrome*, Family Practice 1988:5:4:302 306. pub: Oxford University Press

95. Nairn, C et al. *Comparison of Coxsackie B Neutralisation and Enteroviral PCR in Chronic Fatigue Patients*, Journal of Medical Virology 1995:46:310-313.

96. Peckerman A, LaManca JJ, Dahl KA, Chemitiganti R, Qureishi B, Natelson BH. 2003, *Abnormal Impedance Cardiography Predicts Symptom Severity in Chronic Fatigue Syndrome*, The American Journal of the Medical Sciences: 2003:326:(2):55-60)

97. Pellew RAA, 1955, *A clinical description of a disease resembling poliomyelitis seen in Adelaide*, Med J Aust 1955:42.480 482

98. Poser, Charles , 1992, *'The Differential Diagnosis between Multiple Sclerosis and Chronic Fatigue Postviral Syndrome'* The Clinical and Scientific Basis of Myalgic Encephalomyelitis, Hyde, Byron M.D. (ed) The Nightingale Foundation, Ottawa

99. Preedy VR, Smith DG, Salisbury JR & Peters TJ 1993, *Biochemical and muscle studies in patients with acute onset post-viral fatigue syndrome*, J Clin Pathol. 1993 Aug;46(8):722-6, Department of Clinical Biochemistry, King's College School of Medicine & Dentistry, London

100. Ramsay, A. 1988, *Myalgic Encephalomyelitis and Postviral Fatigue States: The saga of Royal Free Disease*, Gower Medical Publishing, London.

101. Ramsay, AM 1988, *Myalgic encephalomyelitis or what?* Lancet 1988:100

102. Ramsay, AM & Dowsett, EG et al, 1977, *Icelandic Disease (Benign Myalgic Encephalomyelitis or Royal Free Disease)*, BMJ May 1977:1350

103. Ramsay, Melvin A. n.d., *The Myalgic Encephalomyelitis syndrome* [Online], Available: http://www.hfme.org/wramsay.htm

104. Ramsay, Melvin A. 1986, *MYALGIC ENCEPHALOMYELITIS : A Baffling Syndrome With a Tragic Aftermath.* [Online], Available: http://www.hfme.org/wramsay.htm

105. Ramsay, AM. 1956, Encephalomyelitis simulating polio myelitis. Lancet. 1956;1: 761-766

106. Richardson, J. 1999, *Myalgic Encephalomyelitis: Guidelines for doctors* [Online], Available: http://www.hfme.org/wrichardson.htm

107. Richardson, J & Dowsett EG, 1999, *The Epidemiology of Myalgic Encephalomyelitis (ME) in the UK*. Evidence submitted to the All Party Parliamentary Group of Members of Parliament, 23 Nov 1999

108. Richardson, John 2001, *Viral Isolation from Brain in Myalgic Encephalomyelitis (A Case Report), Journal of CFS 2001:9: (3-4):15-19*

109. Richardson, J. n.d., *'M.E., The Epidemiological and Clinical Observations of a Rural Practitioner,'* The Clinical and Scientific Basis of Myalgic Encephalomyelitis, Hyde, Byron M.D. (ed) The Nightingale Foundation, Ottawa, pp. 85-94.

110. Richardson, John 2001, *Enteroviral and Toxin Mediated Myalgic Encephalomyelitis and Other Organ Pathologies*, The Haworth Press Inc. New York

111. Richardson, J. 1995, *DISTURBANCE OF HYPOTHALMIC FUNCTION AND EVIDENCE FOR PERSISTENT ENTEROVIRUS INFECTION IN PATIENTS WITH CHRONIC FATIGUE SYNDROME*, JCFS 1995; 1 (2): 623-624.

112. Richardson. J, Costa D C. 1998, *RELATIONSHIP BETWEEN SPECT SCANS AND BUSPIRONE TESTS IN PATIENTS WITH ME/CFS*, JCFS 1998; 4 (3): 23-38

113. Ryll, Erich D. 1994, INFECTIOUS VENULITIS, CHRONIC FATIGUE SYNDROME, MYALGIC ENCEPHALOMYELITIS, [Online], Available: http://www.geocities.com/tcjrme/recommend23.html

114. Satish R Raj, MD M.S.CI 2006, *The Postural Tachycardia Syndrome (POTS): Pathophysiology, Diagnosis & Management,* Indian Pacing Electrophysiol J. 2006 Apr-Jun; 6(2): 84-99.

115. Sigurdsson B, Sigurjonsson J & Sigurdsson J 1950, *Disease epidemic in Iceland simulating poliomyelitis*, American Journal of Hygiene, 52, 222.

116. Southern P, & Oldstone MBA. 1986, *Medical consequences of persistent viral infection*, New England Journal of Medicine : 1986; 314 : 359-367

117. Streeten, David H. P. 1987, *Orthostatic Disorders of the Circulation, Mechanisms, Manifestations, and Treatment*, Plenum Medical Book Company, New York and London.

118. The Committee for Justice and Recognition of Myalgic Encephalomyelitis 2007, [Online], Available: http://www.geocities.com/tcjrme/

119. Wallis, AL 1957, *An investigation into an unusual disease in epidemic and sporadic form in general practice in Cumberland in 1955 and subsequent years*, University of Edinburgh Doctoral Thesis 1957

120. Williams, Margaret 2007, *Wessely, Woodstock and Warfare*, [Online], Available: http://www.hfme.org.htm

121. Williams, Margaret 2004, *Critical considerations*, [Online], Available: http://www.hfme.org/wmarwillcc.htm

Note that while HFME supports the extremely important work being done by the Nightingale Foundation and Dr Byron Hyde as well as work done by Dr Elizabeth Dowsett and others, views expressed in this book are not their responsibility and are the sole responsibility of the listed authors.

Additional HFME resources available online

Additional HFME papers available on the HFME website include the following:

- A CBT/GET warning letter
- Anaesthesia and M.E.
- Are we just 'marking time?
- A warning on 'CFS' and 'ME/CFS' research and advocacy (co-authored by Lesley Ben)
- M.E. is not fatigue, or 'CFS'
- M.E.: The medical facts: A purely medical overview of the illness including detailed research findings.
- Practical tips for living with M.E.
- Problems with some M.E. (or 'CFS' 'CFIDS' or 'ME/CFS' etc.) advocacy groups
- Problems with the so-called "Fair Name" campaign
- Problems with the use of 'ME/CFS' by M.E. advocates
- Putting research and articles into context
- Smoke and mirrors
- Testing for M.E.
- Testing for M.E.: Plan D
- The effects of CBT and GET on M.E. and The importance of avoiding overexertion in M.E.
- The misdiagnosis of CFS
- The myths about M.E.
- The comprehensive M.E. symptom list
- The WHO ICD in relation to M.E. and 'CFS'
- Treating M.E.: The basics
- What it feels like to have M.E.: A personal M.E. symptom list and description of M.E.
- Who benefits from 'CFS' and 'ME/CFS'?
- Why care about M.E.?
- Why the disease category of 'CFS' must be abandoned
- XMRV, 'CFS' and M.E. by Sarah Shenk. See also: International M.E. expert disputes that 'CFS' XMRV retrovirus claim has relevance to M.E. patients

How to access this information

To view or download any of these additional HFME papers, please visit the www.hfme.org website and click on either the 'Information Guides' or 'Document Downloads' links, or any of the links on the navigation bar. All papers can be downloaded for free from the website in Word or PDF format. To view any of the HFME's web pages or get links for the additional websites listed, please visit the 'Site map' link on the navbar on the www.hfme.org website. New papers are added to the HFME website at least every few months and several new HFME books are being planned for release in 2012 and beyond. To read about new additions to the site, please visit the 'What's new' page on the website or sign up for the free monthly HFME e-newsletter via the website.

'A new idea is first condemned as ridiculous and then dismissed as trivial, until finally, it becomes what everybody knows.'
WILLIAM JAMES (1842-1910)

AFTERWORD

Having Myalgic Encephalomyelitis (M.E). is a traumatic and life-shattering experience and so M.E. patients need all the help they can get. Thank you for taking the time to learn some of the facts of M.E. By putting a little bit of what you have learned into action you can really improve the life of the person you know that has M.E., or the M.E. community generally, which is just wonderful, and no small thing.

I'd like to help further, what should I do?

If you feel you are up to trying to help people with M.E. on a larger scale, please join us in helping to spread the facts about M.E. Unlike people with HIV/AIDS, people with M.E. do not have an initial period of their illness where they are only mildly affected. M.E. is severely disabling even in the first week of illness. People with M.E. are almost all far too ill to stage protests, rallies or marches or even to read about the medical and political facts of the disease.

What we need is people power! Educated people power!

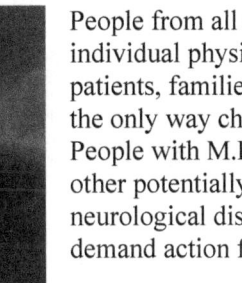

People from all over the world need to stand up for the truth about M.E.; individual physicians, journalists, politicians, human rights campaigners, patients, families and friends of patients, carers and the general public. That is the only way change will occur; through education and insistence on reform. People with M.E. have only a tiny minority of the medical, scientific, legal and other potentially supporting professions on their side. M.E. is an infectious neurological disease that is not going away. We must stand together and demand action from our governments, media and medical bodies.

The insurance companies profiting from 'CFS' are acting without regard for the suffering and avoidable deaths they are causing. These groups are acting criminally. There are powerful and immensely wealthy vested interest groups out there which will continue to fight the truth every step of the way, but we have science, reality and ethics on our side and that is also very powerful. However, for this to be of any use to us, we must first make ourselves aware of the facts and then *use them.* Knowledge is power.

Please help us to spread the truth about Myalgic Encephalomyelitis and expose the lie of 'CFS.' You can also help by NOT supporting the bogus concepts of 'CFS,' 'ME/CFS,' 'subgroups of ME/CFS,' 'CFS/ME,' 'CFIDS' and Myalgic 'Encephalopathy.' Do not give public or financial help or support to groups which promote these harmful and unscientific concepts or which equate M.E. with 'CFS.'

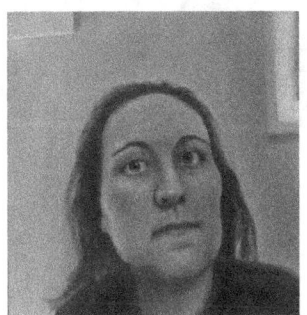

Please see the HFME website for further information on our M.E. advocacy campaigns and all the big and small ways you can help HFME and M.E. patients generally – starting with redistributing HFME leaflets, printouts and books. We're also open to any new ideas you might have. Every contribution helps.

The abuse and neglect of so many severely ill people on such an industrial scale is truly inhuman and has already gone on for far too long. People with M.E. desperately need your help. As anthropologist Margaret Mead famously said, '*Never doubt that a small group of thoughtful, committed citizens can change the world. Indeed, it's the only thing that ever has.*'

Jodi Bassett